SpringerBriefs in Computer Science

SpringerBriefs present concise summaries of cutting-edge research and practical applications across a wide spectrum of fields. Featuring compact volumes of 50 to 125 pages, the series covers a range of content from professional to academic.

Typical topics might include:

- A timely report of state-of-the art analytical techniques
- A bridge between new research results, as published in journal articles, and a contextual literature review
- A snapshot of a hot or emerging topic
- An in-depth case study or clinical example
- A presentation of core concepts that students must understand in order to make independent contributions

Briefs allow authors to present their ideas and readers to absorb them with minimal time investment. Briefs will be published as part of Springer's eBook collection, with millions of users worldwide. In addition, Briefs will be available for individual print and electronic purchase. Briefs are characterized by fast, global electronic dissemination, standard publishing contracts, easy-to-use manuscript preparation and formatting guidelines, and expedited production schedules. We aim for publication 8–12 weeks after acceptance. Both solicited and unsolicited manuscripts are considered for publication in this series.

**Indexing: This series is indexed in Scopus, Ei-Compendex, and zbMATH **

Zhiyuan Wang • Lin Gao • Biying Shou •
Jianwei Huang

Mobile Data Services

Embracing Flexibility in Time, Location, and User Identity

Zhiyuan Wang
School of Cyber Science and Technology
Beihang University
Beijing, China

Biying Shou
School of Management and Economics
The Chinese University of Hong Kong,
Shenzhen
Shenzhen, China

Lin Gao
School of Electronics and Information
Engineering
Harbin Institute of Technology, Shenzhen
Shenzhen, China

Jianwei Huang
School of Science and Engineering
The Chinese University of Hong Kong
Shenzhen, China

Shenzhen Loop Area Institute
Shenzhen, China

ISSN 2191-5768　　　　　　　ISSN 2191-5776　(electronic)
SpringerBriefs in Computer Science
ISBN 978-981-95-1342-0　　　ISBN 978-981-95-1343-7　(eBook)
https://doi.org/10.1007/978-981-95-1343-7

© The Editor(s) (if applicable) and The Author(s), under exclusive license to Springer Nature Singapore Pte Ltd. 2026

This work is subject to copyright. All rights are solely and exclusively licensed by the Publisher, whether the whole or part of the material is concerned, specifically the rights of translation, reprinting, reuse of illustrations, recitation, broadcasting, reproduction on microfilms or in any other physical way, and transmission or information storage and retrieval, electronic adaptation, computer software, or by similar or dissimilar methodology now known or hereafter developed.
The use of general descriptive names, registered names, trademarks, service marks, etc. in this publication does not imply, even in the absence of a specific statement, that such names are exempt from the relevant protective laws and regulations and therefore free for general use.
The publisher, the authors and the editors are safe to assume that the advice and information in this book are believed to be true and accurate at the date of publication. Neither the publisher nor the authors or the editors give a warranty, expressed or implied, with respect to the material contained herein or for any errors or omissions that may have been made. The publisher remains neutral with regard to jurisdictional claims in published maps and institutional affiliations.

This Springer imprint is published by the registered company Springer Nature Singapore Pte Ltd.
The registered company address is: 152 Beach Road, #21-01/04 Gateway East, Singapore 189721, Singapore

If disposing of this product, please recycle the paper.

To everyone dissatisfied with the past mobile data services—your frustration inspired this work.

Preface

Mobile Network Operators (MNOs) monetize the scarce wireless spectrum resource, generating revenue through the provision of mobile data services to their customers. The traditional mobile data service is a three-part tariff involving a monthly data quota. This kind of service lacks for flexibility seriously, since it rigidly dictates the usage of data based on time, location, and user identity, limiting when, where, and by whom the data can be consumed. The intensifying market competition, driven by 5G deployment and the proliferation of IoT devices, has recently forced MNOs to explore more flexible data services. For instance, the rollover data service enables time flexibility by permitting unused data in the current month to be carried over for usage in the subsequent month. The day-pass data service enables location flexibility by granting MUs the ability to utilize their domestic data while traveling overseas. The data trading service enables user-identity flexibility by creating a marketplace where MUs can either offload their surplus data or purchase additional data from one another.

In this book, we delve into the economic issues of flexible mobile data services. We will analyze MUs' optimal behaviors and MNOs' data service optimization. The book aims to reveal who ultimately reaps the benefits from flexible mobile data services—whether it will be the mobile MUs or the MNOs themselves.

We start in Chap. 1 by discussing the motivation for us to write (and for you to read) this book. Specifically, we will introduce the traditional data service and its drawbacks in telecommunication market. We then introduce how industry (e.g., AT&T) improves their data services by enhancing the flexibility from time, location, and user dimensions, and review the related studies.

In Chaps. 2 and 3, we focus on rollover data service with time flexibility. Specifically, Chap. 2 considers a monopoly mobile data market. We will analyze MUs' subscription behavior and the MNO's rollover data service optimization. We will also introduce how the MNO proceeds price discrimination to meet the needs of different MUs. This chapter will unveil the economic impact of time flexibility on the monopoly MNO's profit and the MUs' payoffs. Chapter 3 considers a competitive mobile data market, where multiple MNOs compete for market share. We will analyze the MNOs' pricing equilibrium and the data service adoption equilibrium. This chapter will answer whether all the MNOs should enhance the time flexibility by adopting rollover data service.

In Chap. 4, we move on to the day-pass data service with location flexibility. Specifically, we will analyze the MU's optimal data consumption and day-pass service configuration based on the location user profile. Moreover, we will also investigate how the MNO monetizes its day-pass service with the uncertainty of user demand.

In Chaps. 5 and 6, we focus on the data trading service with user-identity flexibility. Specifically, in Chap. 5, we will analyze mobile MUs' optimal trading strategy and the MNO's revenue-maximizing trading prices. Moreover, we will also investigate how the rollover data service affects the equilibrium between mobile MUs and the MNO. This chapter will unveil the economic viability of integrating time flexibility and user flexibility. In Chap. 6, we focus on the integrated time-user flexibility and further investigate how bounded rational MUs behave to it. This chapter will unveil the impact of user bounded rationality on the economic benefit of flexible data services.

Finally, in Chap. 7, we conclude the flexible mobile data services studied in this book and provide an outlook of some open issues and the future challenges in the telecommunication market.

As mobile networks transition toward 6G and integrate artificial intelligence for network optimization, the flexible data services studied in this book become increasingly relevant for managing heterogeneous traffic patterns and diverse user requirements in next-generation wireless systems.

Beijing, China	Zhiyuan Wang
Shenzhen, China	Lin Gao
Shenzhen, China	Biying Shou
Shenzhen, China	Jianwei Huang
July 2025	

Acknowledgment We extend our sincere gratitude to the researchers at the Network Communications and Economics Laboratory (NCEL), part of the Department of Information Engineering at the Chinese University of Hong Kong. Their insightful and valuable feedback played a crucial role in shaping the research papers that serve as the foundation for this book. Zhiyuan Wang would like to acknowledge the support received from the National Natural Science Foundation of China (Grant U24B20128 and Grant 62202021). Lin Gao would like to acknowledge the support received from the Natural Science Foundation of Guangdong Province under Grant 2024A1515010178, Shenzhen Science and Technology Program under Grant KQTD20190929172545139, Grant GXWD20231129103946001, Grant KJZD20240903095402004, and Grant ZDSYS20210623091808025. Jianwei Huang would like to acknowledge the support received from the National Natural Science Foundation of China (Project 62271434), the Shenzhen Stability Science Program 2023, the Shenzhen Institute of Artificial Intelligence and Robotics for Society, and Longgang District Shenzhen's "Ten Action Plan" for Supporting Innovation Projects (No. LGKCSDPT2024002).

Contents

1 **Mobile Data Services** 1
   1.1 Background 1
   1.2 Innovations on Flexible Data Services 2
       1.2.1 Rollover Data Service 2
       1.2.2 Day-Pass Service 3
       1.2.3 Data Trading Service 3
   1.3 Related Studies on Mobile Data Services 4
       1.3.1 Traditional Three-Part Tariff 4
       1.3.2 Rollover Data Plan 4
       1.3.3 Data Trading Market 4
       1.3.4 Location-Dependent Data Services 5
   1.4 Road-Map of This Book 5

2 **Time Flexibility in Monopoly Market** 7
   2.1 Market Model 7
       2.1.1 Mobile Data Services 8
       2.1.2 MU Model 9
       2.1.3 MNO Model 10
   2.2 Degree of Time Flexibility 11
   2.3 User Subscription 14
   2.4 MNO's Data Service Optimization 16
       2.4.1 Optimal Pricing Strategy 16
       2.4.2 Optimal Data Cap 17
       2.4.3 Optimal Data Mechanism 18
   2.5 MNO's Price Discrimination 18
       2.5.1 Contract Formulation 18
       2.5.2 Marginal Rate of Substitution 21
       2.5.3 Necessary Conditions for Feasible Contract 23
       2.5.4 Sufficient Conditions for Feasible Contract 24

		2.5.5	Optimal Contract Design	26
		2.5.6	Dynamic Quota Allocation (DQA) Algorithm	28
	2.6	Summary		30
3	**Time Flexibility in Competitive Market**			31
	3.1	Market Model		31
	3.2	User Subscription		32
	3.3	MNOs' Pricing Competition		34
		3.3.1	Best Response Analysis	35
		3.3.2	Equilibrium Analysis	38
	3.4	MNOs' Mechanism Competition		39
		3.4.1	Market Partition at Game 3.2 Equilibrium	39
		3.4.2	Single-MNO-Surviving	40
		3.4.3	Coexistence	41
	3.5	Summary		42
4	**Location Flexibility in Overseas Market**			43
	4.1	Market Model		43
		4.1.1	Mobile Data Services	43
		4.1.2	User Model	44
		4.1.3	MNO Revenue	46
	4.2	Off-line Solution and Insights		46
		4.2.1	Off-line Problem and Reformulation	47
		4.2.2	Key Insights of Solving Problem 4.3	48
		4.2.3	Solution of Problem 4.3	51
	4.3	MU's Online Strategy		52
	4.4	Summary		56
5	**User-identity Flexibility in Data-trading Market**			59
	5.1	Market Model		59
		5.1.1	Wireless Data Services	59
		5.1.2	MUs' Decisions	60
		5.1.3	MNO's Decision	64
	5.2	MU's Trading Policy		65
		5.2.1	Plain Trading	65
		5.2.2	Rollover-involved Trading	66
		5.2.3	Impact of Rollover Mechanism	67
	5.3	MNO's Optimal Pricing		68
	5.4	Summary		70
6	**Interplay between Time and User-identity Flexibility**			71
	6.1	Market Model		71
		6.1.1	User Model	72
		6.1.2	User's Consumption and Trading Problem	74
		6.1.3	Trading Completion Ratio	76

6.2	User's Consumption and Trading Policy		77
	6.2.1	Optimal Consumption and Trading Decisions	77
	6.2.2	Expected Monthly Payoff	80
	6.2.3	User Subscription	86
6.3	MNO's Optimal Pricing		87
	6.3.1	MNO's Revenue	88
	6.3.2	MNO's Pricing Problem	88
	6.3.3	Pricing Without Buying Service	89
	6.3.4	Pricing with Buying Service	91
	6.3.5	Summary	93
6.4	Discussion and Extension		95
6.5	Summary		96

7 Conclusion and Outlook ... 97
   7.1 Conclusion ... 97
   7.2 Outlook ... 98
      7.2.1 General Competitive Market ... 98
      7.2.2 Multi-Dimensional Flexibility ... 99
      7.2.3 Bounded Rationality ... 100

References ... 101

Acronyms

CapEx	Capital Expenditure
IC	Incentive Compatibility
IFR	Increasing Failure Rate
IR	Individual Rationality
MNO	Mobile Network Operator
MRS	Marginal Rate of Substitution
MU	Mobile User
OpEx	Operation Expenditure
PIC	Pairwise Incentive Compatibility

Chapter 1
Mobile Data Services

Abstract This chapter introduces the motivation for us to write (and for you to read) this book. Specifically, we will introduce the traditional mobile data service and its drawbacks in telecommunication market. We then introduce how industry operators (e.g., AT&T) improves their data services by enhancing the flexibility from time, location, and user-identity dimensions, and review the related studies.

Keywords Mobile network · Mobile data plans · Data pricing · Rollover data · Data trading · AT&T · China Mobile

1.1 Background

In the telecommunication market, Mobile Network Operators (MNOs) profit from providing mobile data services through carefully designing and optimizing their data plans. The most widely adopted data plan used to be the flat rate scheme, where the mobile user (MU) pays a monthly fee for the unlimited data usage. However, the MNOs have been suffering from the network congestion due to the explosive growth of the mobile data traffic. According to Ericsson's report [1], the global monthly mobile data traffic has grown 14-fold from 2017 to 2024, and reaches more than 140 exabytes (i.e., 140 billion billion bytes) per month. This challenge is further amplified by emerging technologies such as autonomous vehicles, smart cities, and AI-powered applications that require adaptive data allocation strategies. The explosive growth in mobile data traffic has forced most MNOs to introduce the monthly data cap to alleviate the network congestion. Nowadays the most common data plan is a three-part tariff that consists of a data cap, a subscription fee, and a per-unit overage fee (for exceeding the data cap). Depending on the realized data demand, subscribers face two possible scenarios:

- In a light-demand month, the leftover data cap expires at the end of this month.
- In a heavy-demand month, the data cap soon runs out and the speed of mobile network service will be reduced by the MNO, unless the user purchases additional common-speed data.

The introduction of data caps, while intended to alleviate network congestion, inadvertently contributes to inefficiencies within the mobile data market. According to the research conducted by [2], on average each MU in UK has 3.4GB of paid but unused mobile data every month. This kind of data service lacks for flexibility seriously, since it rigidly dictates the usage of data based on time, location, and user identity, limiting when, where, and by whom the mobile data cap can be consumed. To sum up, the traditional mobile data service hurts both the MU's payoffs and the MNO's revenue, leading to a market with the poor welfare.

To address the aforementioned drawbacks, there is a prevailing hope that MNOs will eliminate data caps by adopting advanced technologies to enhance the network capacity. However, such technological upgrades are often expensive and time-intensive. When the authors write this book, AT&T have been offering so-called "unlimited plans". However, the premium speed is only guaranteed within the data allowance (e.g., 50GB or 100GB). Hence it is actually a special three-part tariff plan. **An economic solution is to increase the flexibility of the fixed data cap from different dimensions**.

- **Time Flexibility**: The rollover data service enables time flexibility by permitting the unused data in the current month to be carried over for usage to the subsequent month(s).
- **Location Flexibility**: The day-pass data service enables location flexibility by granting MUs the ability to utilize their domestic data cap while they are traveling overseas.
- **User-identity Flexibility**: The data trading service enables user-identity flexibility by creating a marketplace where MUs can either offload their surplus data cap or purchase additional data from one another. Particularly, the MNOs may also take part in data trading as "special MUs".

1.2 Innovations on Flexible Data Services

In the following, we introduce some flexible data services that have been experimented by industry MNOs for several years all over the world.

1.2.1 Rollover Data Service

Rollover data services vary primarily in two aspects: (1) the consumption priority between rollover data (carried over from the previous month) and the current monthly data allowance, and (2) the duration for which unused data remains valid. These factors determine the extent to which a user's effective data cap can be increased. Under a traditional data plan, any unused data expires at the end of the billing cycle. In contrast, AT&T's rollover service [3] permits subscribers to utilize

1.2 Innovations on Flexible Data Services 3

leftover data from the previous month, but only within the current month-expiring thereafter. Notably, AT&T's policy stipulates that rollover data is only used after the current month's primary data allowance is exhausted. Similarly, China Mobile [4] also allows one-month rollover, but with a key distinction: the rollover data is consumed before dipping into the current month's allocated data. Like AT&T, unused rollover data expires after the subsequent billing cycle. A more flexible approach is adopted by Sky Mobile [5], which lets users accumulate unused data in a "Sky Piggybank", enabling longer-term savings compared to the single-month rollover policies of AT&T and China Mobile.

1.2.2 Day-Pass Service

Several MNOs, including AT&T and China Mobile Hong Kong (CMHK), have recently launched a location-flexible data service to address the limitations of location-restricted data plans. The core concept allows MUs to use their existing data plans across different locations on demand (as described in AT&T's Passport service [6]). Instead of solely relying on costly roaming add-ons when traveling abroad, AT&T subscribers now have an alternative: enabling location flexibility for a daily fee (e.g., $10 per day) to access their monthly data allowance. Under this model, any data used on a given day is deducted from the subscriber's monthly quota. While this service offers clear advantages for travelers, its effectiveness depends on the MU's data usage patterns. Since the monthly data cap must now cover both domestic and international usage, enabling location flexibility could lead to additional overage charges if the allocated data is exhausted prematurely.

1.2.3 Data Trading Service

The adoption of data trading services differs in the data supply pattern and the minimal trading amount. First, the data supply pattern determines whether the MUs can buy extra or sell leftover data. For example, both AT&T [3] and China Mobile [4] allow subscribers to buy extra common-speed data. SMARTY [7] gives MUs money for data that they have not used at the end of each month. Moreover, CMHK [8] has been operating a data trading platform called "2CM", where MUs can buy and sell data with each other. Second, the minimal trading amount determines how flexible the supply pattern can be. For example, the minimal trading amount specified by CMHK [8] is 1GB, which may impede the MUs' optimal trading decisions. In contrast, China Mobile offers a more convenient choice by adopting a minimal amount of 100MB. Moreover, SMARTY [7] has an even more aggressive policy and allows MUs to sell data down to the last 1MB.

1.3 Related Studies on Mobile Data Services

There have been many excellent studies on mobile data services and pricing. We refer interested readers to the comprehensive survey [9, 10]. In the following, we briefly introduce the studies that are closely related to this book.

1.3.1 Traditional Three-Part Tariff

The traditional mobile data service is a three-part tariff, consisting of a fixed fee, an allowance of free units, and a price per unit above the allowance [11]. Previous studies focused on MUs' data consumption behaviors and the optimal contract design. For example, Lambrecht et al. in [12] conducted an empirical study on MUs' data consumption behaviors and assessed how the consumer's demand uncertainty affects the tariff choices. Xu et al. in [13] studied how individual MUs dynamically consume the data allowance under the three-part tariff scheme, and identified the characteristics of forward-looking and myopic MUs. Nevo et al. in [14] focused on the variation in the shadow price of usage under the three-part tariff, and provided evidence that the MUs will respond to this variation. Fibich et al. in [15] discussed the challenge of solving the optimal three-part tariff, and explicitly characterized the optimal three-part tariff plan. Bhargava et al. in [16] found that the optimal three-part tariff outcomes are identical to the optimal two-part tariff outcomes, when the market demand follows an increasing price elasticity. Wu et al. in [17] solved the optimal three-part tariff design problem with multiple consumer types.

1.3.2 Rollover Data Plan

The rollover data service was just proposed in recent years, which has not been extensively studied yet. As far as we know, Zheng et al. in [18] provided the first study on data rollover service. They focused on the data rollover service offered by AT&T, and overlooked the impact of the consumption priority on the MNO and MUs. Wang et al. in [19] studied the optimization of the MNO's data plan with rolling service and found that the consumption flexibility can increase both MNO's profit and all MUs' expected payoff. Wei et al. in [20] studied the rollover period length from a profit-maximizing MNO's perspective. However, the above studies only characterized the MUs through the random data demand without considering the MUs' dynamic data consumption behaviors.

1.3.3 Data Trading Market

Data trading service (including both buying and selling choices) has not been well studied before. Zheng et al. in [21] purely examined the data trading market of CMHK, where MUs trade data with each others by submitting their bids.

They derived MUs' optimal behavior and proposed an algorithm for the revenue-maximizing MNO to match buyers and sellers. Yu et al. in [22] focused on the MUs' behaviors faced with demand uncertainty based on prospect theory. They found that a risk-averse dominant MU guarantees a higher minimum profit, while a risk-seeking dominant MU achieves a higher maximum profit. Huang et al. in [23] found that data trading among MUs can enhance the service provider's profitability, since data trading address MUs' overage disutility.

1.3.4 Location-Dependent Data Services

Several studies have explored how location-dependent pricing can enhance MNO's revenue streams. Ma et al. [24] introduced a time-and-location-aware pricing model to incentivize MUs to distribute network traffic more evenly, thereby alleviating congestion. Maille et al. [25] analyzed how usage-based roaming charges impact MNO profitability. Duan et al. [26] evaluated third-party roaming data services under both flat-rate and usage-based pricing schemes. Wang et al. [27] investigated cooperative roaming via user-generated hotspots. Zhang et al. [28] examined hybrid pricing models for roaming data access in both competitive and collaborative market environments. However, despite these advancements, location-flexible data services—where MUs can dynamically apply their domestic data allowances abroad—-remain unexplored in existing literature to the best of our knowledge.

1.4 Road-Map of This Book

In this book, we will delve into the economic issues of flexible mobile data services. We will leverage game theory to analyze MUs' optimal behavior, and also design economic mechanisms for MNOs' data service optimization. This book aims to reveal who ultimately reaps the benefits from flexible mobile data services—whether it will be the MUs or the MNOs themselves. Part of the results in this book are based on the previous studies in [29–37]. Chapters 2 and 3 focus on the rollover data service with time flexibility in a monopoly market and a competitive market, respectively. In Chap. 4, we move on to the day-pass data service with location flexibility. In Chap. 5, we focus on data trading service with user-identity flexibility, and analyze MUs' optimal trading strategy and the MNO's revenue-maximizing trading prices. In Chap. 6, we focus on the interplay between time flexibility and user-identity flexibility. We believe that this book could provide a comprehensive understanding on the benefits of flexible mobile data services, and facilitate the efficient operation of practical telecommunication market.

Chapter 2
Time Flexibility in Monopoly Market

Abstract This chapter considers a monopoly mobile data market. We will analyze MUs' subscription behavior and the MNO's rollover data service optimization. We will also introduce how the MNO proceeds price discrimination to meet the needs of different MUs. This chapter will unveil the economic impact of time flexibility on the monopoly MNO's profit and the MUs' payoffs.

Keywords Rollover data service · Time flexibility · Contract theory · Monopoly market

2.1 Market Model

We analyze a monopoly telecommunication market where a single Mobile Network Operator (MNO) provides mobile data services to heterogeneous mobile users (MUs). The MNO designs a data plan to maximize its profit, while each MU independently decides whether to subscribe based on their own payoff maximization. We use a three-stage Stackelberg game formulation to characterize how the MNO and MUs interact with each other. Specifically, the MNO act as the Stackelberg leader, and the MUs act as the followers.

- Stage I: MNO selects the data mechanism for its three-part tariff plan.
- Stage II: MNO determines the data cap, subscription fee, and per-unit fee.
- Stage III: MUs make their subscription decisions to maximize individual payoffs.

Next, we first build a unified framework for mobile data plans with/without time flexibility, and then present the game-theoretic model in details from the perspectives of both the MUs and the MNO.

2.1.1 Mobile Data Services

We model the mobile data services as a tuple $\mathcal{T} = \{Q, \Pi, \pi, \kappa\}$. Specifically, the subscriber pays a fixed lump-sum subscription fee Π (e.g., in USD) for his/her monthly data consumption up to the data cap Q (e.g., in GB). The subscriber has to pay the overage fee π (e.g., in USD/GB) for each unit of additional data consumption (exceeding the data cap). The parameter κ here is used to represent *data mechanisms*. Subscribing to different data mechanisms, the subscribers may enjoy different degrees of time flexibility.

This chapter focuses on three data mechanisms, and we use $\kappa \in \{0, 1, 2\}$ to index them. The key difference among the three data mechanisms is twofold, i.e., special data and its consumption priority. In general, special data is actually the rollover data that the subscriber obtained from the leftover data cap of previous month(s). This kind of special data will enlarge the MU's effective data cap (within which the subscriber does not need to pay overage fee) in the current month. In particular, how large the effective data cap becomes not only depends on the existence of special data, but also the consumption priority between the special data and the current monthly data cap. For illustration purpose, Table 2.1 shows the key differences among the three data mechanisms. Mathematically, τ represents an MU's data surplus at the beginning of a month, and $Q_\kappa^e(\tau)$ denotes the corresponding effective data cap given the data mechanism κ.

- The case of $\kappa = 0$ represents the **traditional data mechanism**. Its subscribers have no special data. The effective data cap is $Q_0^e(\tau) = Q$.
- The case of $\kappa = 1$ represents the **rollover data mechanism** of AT&T [3]. The rollover data τ carried from the previous month will be used *after* the current monthly data cap. Hence, the effective cap is $Q_1^e(\tau) = Q + \tau$.
- The case of $\kappa = 2$ represents the **rollover data mechanism** of China Mobile [4]. The rollover data τ carried from the previous month is used *prior* to the current monthly data cap. Hence, the effective data cap is $Q_2^e(\tau) = Q + \tau$.

As one can imagine, time flexibility essentially enlarges the MU's effective data cap. As shown in Table 2.1, the effective data cap of the traditional data mechanism (i.e., $\kappa = 0$) is always Q. Nevertheless, the potential maximal value of the effective data cap is $2Q$ for the rollover mechanisms (i.e., $\kappa = 1$ and $\kappa = 2$). The greater the effective data cap becomes, the MUs incur less overage usage. This eventually changes the MUs' subscription decisions in the telecommunication market.

Table 2.1 Mobile data services $\mathcal{T} = \{Q, \Pi, \pi, \kappa\}$

Mechanism	Data surplus	Consumption priority	Effective data cap
$\kappa = 0$	None	/	Q
$\kappa = 1$	Rollover data τ	Data cap $\Rightarrow$ Rollover data	$Q + \tau$
$\kappa = 2$	Róllover data τ	Rollover data $\Rightarrow$ Data cap	$Q + \tau$

2.1.2 MU Model

MU Characteristics We model the generic MU from three aspects, i.e., the demand for mobile data, the valuation for mobile data, and the substitutability for mobile network. First of all, data demand is usually measured in the minimum data unit (e.g, 1KB or 1MB according to the MNO's billing practice). Hence, the MU's monthly data demand is modeled as a discrete random variable d. We let $f(d)$ denote the probability mass function on the support $\{0, 1, 2, \ldots, D\}$. Some studies (e.g., [12, 14]) show that d approximately follows a truncated log-normal distribution. Second, we follow the previous studies (e.g., [38]) and denote θ as the MU's valuation (i.e., the utility) from consuming one unit of data. It is usually the private information that the MNO cannot obtain exactly. Third, we would like to explore the behavior change of the MU when her/his data consumption exceeds the effective data cap $Q_\kappa^e(\tau)$. This is because further data consumption will cause extra payment for the MUs. In this book, we suppose that the MU will continue to use the Internet services and consume data, but she/he would voluntarily rely more heavily on other alternative networks (such as Wi-Fi). To this end, we follow previous studies (e.g., [39]) and use the network substitutability $\beta \in [0, 1]$ to denote the fraction of overage usage shrink. Although the MU population may have other proprieties, this book will focus on the heterogeneity regarding the data valuation θ and the network substitutability β. Accordingly, each MU is characterized by (θ, β), and the whole MU market is mathematically defined as follows:

$$\mathcal{M} \triangleq \{(\theta, \beta) : 0 \leq \theta \leq \theta_{\max}, 0 \leq \beta \leq 1\}. \tag{2.1}$$

MU Payoff To model the economic behavior of MUs, we define the MU's payoff as the utility (of consuming mobile data) minus the total payment. Consider a type-(θ, β) MU who has d units of mobile data demand and the effective data cap $Q_\kappa^e(\tau)$. The alternative network allows the MU to reduce the overage data demand, thus the data usage of this MU is actually $d - \beta[d - Q_\kappa^e(\tau)]^+$. Here we have $[x]^+ = \max\{0, x\}$. Obviously, the overage data usage of this MU is given by $(1 - \beta)[d - Q_\kappa^e(\tau)]^+$. Furthermore, we denote ρ as the MNO's quality of service (QoS), which affects subscribers' quality of experience (QoE) when using the mobile data services. Roughly speaking, we use ρ as a multiplicative coefficient of MU's utility. Accordingly, the MU's utility of consuming $d - \beta[d - Q_\kappa^e(\tau)]^+$ units of mobile data is given by $\rho\theta(d - \beta[d - Q_\kappa^e(\tau)]^+)$. Regarding to MU's payment, there are two aspects, consisting of the subscription fee Π and the overage payment. Specifically, the overage data usage $(1 - \beta)[d - Q_\kappa^e(\tau)]^+$ will lead to overage payment $\pi(1 - \beta)[d - Q_\kappa^e(\tau)]^+$. For a type-$(\theta, \beta)$ MU, if the data demand is d and the effective cap is $Q_\kappa^e(\tau)$, then the payoff is given by

$$S(\mathcal{T}, \theta, \beta, d, \tau) = \theta\rho\left(d - \beta[d - Q_\kappa^e(\tau)]^+\right) \\ - \pi[d - Q_\kappa^e(\tau)]^+(1 - \beta) - \Pi. \tag{2.2}$$

Note that d and τ are two random variables. We take the expectation over d and τ, and then obtain a type-(θ, β) MU's expected monthly payoff as

$$\begin{aligned}\bar{S}(\mathcal{T}, \theta, \beta) &= \mathbb{E}_{d,\tau}\left[S\left(\mathcal{T}, \theta, \beta, d, \tau\right)\right] \\ &= \theta\rho\left[\bar{d} - \beta A_\kappa(Q)\right] - \pi A_\kappa(Q)(1 - \beta) - \Pi.\end{aligned} \quad (2.3)$$

In (2.3), $A_\kappa(Q)$ represents the type-$(\theta, 0)$ MU's expected monthly overage data consumption, which depends on the data mechanism κ. Mathematically, it is defined as follows:

$$\begin{aligned}A_\kappa(Q) &= \mathbb{E}_{d,\tau}\left\{\left[d - Q_\kappa^e(\tau)\right]^+\right\} \\ &= \sum_\tau \sum_d [d - Q_\kappa^e(\tau)]^+ f(d) p_\kappa(\tau).\end{aligned} \quad (2.4)$$

In (2.4), $p_\kappa(\tau)$ is the probability mass function (PMF) of τ under the data mechanism κ. As we will see later, the differences among the three data mechanisms lie in this function. We will explain how to mathematically derive $p_\kappa(\tau)$ and $A_\kappa(Q)$ in details later. Now we move on to the entire MU population, and compute the MNO's total payoff as follows:

$$\tilde{S}(\mathcal{T}) = \iint_{\Psi(\mathcal{T})} \bar{S}(\mathcal{T}, \theta, \beta) h(\theta) g(\beta) d\theta d\beta. \quad (2.5)$$

2.1.3 MNO Model

MNO Revenue The MNO's revenue obtained from a single MU includes the subscription fee and the possibly overage fee. Therefore, the *MNO's revenue from a type-(θ, β) subscriber* with d units of data demand and an effective cap $Q_\kappa^e(\tau)$ is

$$R(\mathcal{T}, \theta, \beta, d, \tau) = \pi(1 - \beta)\left[d - Q_\kappa^e(\tau)\right]^+ + \Pi. \quad (2.6)$$

Recall that there are two random variables (i.e., d and τ). Hence we further take the expectation and derive the MNO's expected monthly revenue from a type-(θ, β) MU as follows:

$$\begin{aligned}\bar{R}(\mathcal{T}, \theta, \beta) &= \mathbb{E}_{d,\tau}\left[R\left(\mathcal{T}, \theta, \beta, d, \tau\right)\right] \\ &= \pi(1 - \beta) A_\kappa(Q) + \Pi,\end{aligned} \quad (2.7)$$

2.2 Degree of Time Flexibility

Moving on to the entire MU population, we obtain the MNO's expected total revenue from the entire market as follows:

$$\tilde{R}(\mathcal{T}) = \iint_{\Psi(\mathcal{T})} \bar{R}(\mathcal{T}, \theta, \beta) h(\theta) g(\beta) \mathrm{d}\theta \mathrm{d}\beta. \tag{2.8}$$

MNO Cost This book focuses on two types of typical costs perceived by the MNO, i.e., capacity cost and operational cost. To be more specific, the capacity cost is a cost that has already been incurred and cannot be recovered, while the operational cost is a cost that may be incurred or changed if an action is taken [40]. First, the MNO's capacity cost is primarily driven by capital expenditures (CapEx), which reflect investments in network infrastructure [41]. In practice, data caps serve as a tool for MNOs to control network congestion and allocate limited bandwidth efficiently [42]. Consequently, many MNOs implement data caps to mitigate congestion-related issues [43]. Thus, when an MNO introduces a data cap, it must ensure sufficient network capacity is available to handle the anticipated traffic. Building on this insight, we suppose that the MNO's capacity expenditure or cost as an increasing function of the data cap. Intuitively, higher data caps lead to greater network strain, necessitating larger upfront investments in capacity. Second, the MNO's operational expenses (OpEx) stem mainly from system management and maintenance [44]. Once a mobile data plan is deployed, the MNO's OpEx is influenced by subscribers' aggregate data usage. Specifically, the expected total data consumption is

$$L(\mathcal{T}) = \iint_{\Psi(\mathcal{T})} \left[\bar{d} - \beta A_\kappa(Q) \right] h(\theta) g(\beta) \mathrm{d}\theta \mathrm{d}\beta. \tag{2.9}$$

We follow previous studies (e.g., [45]) and adopt a linear operational cost model $L(\mathcal{T}) \cdot c$, where c denotes the marginal operational cost from unit data consumption. Combining both capacity and operational costs, we express the MNO's expected total cost for a given plan as:

$$\tilde{C}(\mathcal{T}) = L(\mathcal{T}) \cdot c + J(Q). \tag{2.10}$$

MNO Profit Given the above discussion, the MNO's profit is formally defined as the difference between revenue and cost. The MNO's expected profit under $\mathcal{T}$ is

$$\tilde{W}(\mathcal{T}) = \tilde{R}(\mathcal{T}) - \tilde{C}(\mathcal{T}). \tag{2.11}$$

2.2 Degree of Time Flexibility

As mentioned in Sect. 2.1.2, the three data mechanisms differ in the data surplus. More specifically, it is the distribution of the MU's data surplus that matters, and such a distribution will further affect $A_\kappa(Q)$. Particularly, the data surplus of

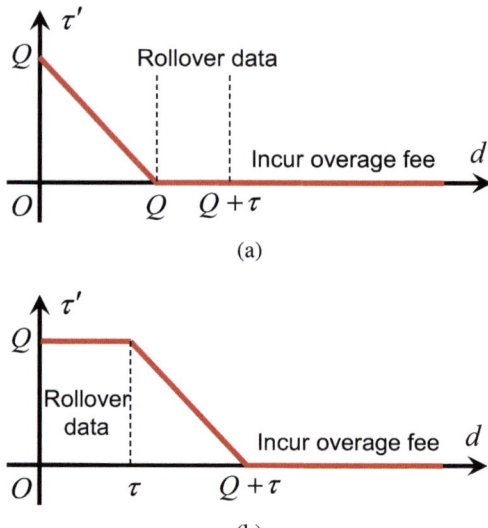

Fig. 2.1 Illustration of the transition of rollover data under different rollover mechanisms. (**a**) $\kappa = 1$. (**b**) $\kappa = 2$

traditional data mechanism $\kappa = 0$ is always zero (i.e., $\tau = 0$). For the other rollover mechanisms $\kappa \in \{1, 2\}$, it is necessary to investigate the transition between successive months. We use Fig. 2.1 to showcase this transition. The horizontal axis and vertical axis represent MUs' data demand d and MUs' data surplus τ' for the next month, respectively. In this case, the differences between the red curves in Fig. 2.1 showcase how the two data mechanisms $\kappa \in \{1, 2\}$ differ with each other. Next we introduce $p_\kappa(\tau)$ and $A_\kappa(Q)$ given the data mechanisms.

Traditional Data Mechanism $\kappa = 0$ Consider a generic $\mathcal{T} = \{Q, \Pi, \pi, 0\}$ subscriber. No special data is available. That is, the random variable $\tau = 0$ and we have $Q_0^e(\tau) = Q$. In this case, we only need to consider the randomness from data demand. Therefore, $A_0(Q)$ is

$$A_0(Q) = \sum_{d=0}^{D} [d - Q]^+ f(d). \tag{2.12}$$

Rollover Data Mechanism $\kappa = 1$ Consider a generic MU subscribing to $\mathcal{T} = \{Q, \Pi, \pi, 1\}$. He/She has rollover data from the previous month, which is used after the monthly data cap. The effective data cap is $Q_1^e(\tau) = Q + \tau$. Figure 2.1a shows the rollover data (to the next month) versus the MU's data demand (in the current month). We have

$$\tau' = \begin{cases} 0, & \text{if } d \geq Q, \\ Q - d, & \text{if } d < Q. \end{cases} \tag{2.13}$$

2.2 Degree of Time Flexibility

Note that τ' depends on Q and d, but is independent of τ. The probability mass function $p_1(\tau)$ is given by

$$p_1(\tau) = \begin{cases} \sum_{d=Q}^{D} f(d), & \text{if } \tau = 0, \\ f(Q - \tau), & \text{if } \tau \in (0, Q]. \end{cases} \quad (2.14)$$

Now we take the expectation over d and τ to compute the expected overage usage $A_1(Q)$, as follows:

$$A_1(Q) = \sum_{\tau=0}^{Q} \sum_{d=0}^{D} [d - Q_1^e(\tau)]^+ f(d) p_1(\tau). \quad (2.15)$$

Rollover Data Mechanism $\kappa = 2$ Consider a generic MU subscribing to $\mathcal{T} = \{Q, \Pi, \pi, 2\}$. He/She has rollover data carried over from the previous month. It is used prior to the current monthly data cap. The effective data cap is $Q_2^e(\tau) = Q + \tau$. Recall that the rollover data is consumed *prior* to the monthly cap. According to the transition shown in Fig. 2.1b, we obtain

$$\tau' = \begin{cases} Q, & \text{if } d \in [0, \tau], \\ Q + \tau - d, & \text{if } d \in (\tau, Q + \tau), \\ 0, & \text{if } d \in [Q + \tau, D]. \end{cases} \quad (2.16)$$

Note that τ' depends on Q, d, and τ. This leads to the Markov property over τ. The one-step transition probability is

$$p_2(\tau, \tau') = \begin{cases} \sum_{d=0}^{\tau} f(d), & \text{if } \tau' = Q, \\ f(Q + \tau - \tau'), & \text{if } \tau' \in (0, Q), \\ \sum_{d=Q+\tau}^{D} f(d), & \text{if } \tau' = 0. \end{cases} \quad (2.17)$$

One could numerically compute the stationary distribution of the rollover data surplus τ, denoted by $p_2(\tau)$, according to the above transition probability [46]. Accordingly, $A_2(Q)$ is

$$A_2(Q) = \sum_{\tau=0}^{Q} \sum_{d=0}^{D} [d - Q_2^e(\tau)]^+ f(d) p_2(\tau). \quad (2.18)$$

So far, we have derived $A_\kappa(Q)$ under the three data mechanisms. In the following, we further evaluate and compare the degree of time flexibility of the three data mechanisms based on $A_\kappa(Q)$. As shown in Definition 2.1, we measure time flexibility based on the type-$(\theta, 0)$ MU's expected overage data consumption.

Definition 2.1 For any data mechanisms $i, j \in \{0, 1, 2\}$, we say that i has a better time flexibility than j (which is denoted by $\mathcal{F}_i > \mathcal{F}_j$) if and only if we have $A_i(Q) < A_j(Q)$ no matter how the data demand distribution varies.

The intuition behind Definition 2.1 is that a better time flexibility should lead to less overage usage for its subscribers under the same data cap Q for arbitrary demand distributions. Lemma 2.1 summarizes the results.

Lemma 2.1 *For any demand distribution $f(\cdot)$, we have*

$$A_0(Q) > A_1(Q) > A_2(Q), \quad \forall Q \in (0, D). \tag{2.19}$$

In other words, the time flexibility offered by the three data mechanisms satisfies the following conditions

$$\mathcal{F}_0 < \mathcal{F}_1 < \mathcal{F}_2. \tag{2.20}$$

Lemma 2.1 shows that the rollover data mechanism $\kappa = 2$ (China Mobile) provides the best time flexibility. The key reason why $\mathcal{F}_1 < \mathcal{F}_2$ lies in the irregular consumption priority under $\kappa = 1$ (AT&T's rollover mechanism). Specifically, AT&T's mechanism requires subscribers to exhaust their current monthly data cap (Q) before using any rollover data from the previous month. This means later-allocated data (current month) takes precedence over earlier-allocated data (rollover), disrupting natural usage patterns. As a result, subscribers cannot fully optimize their long-term data utilization, reducing effective time flexibility. Despite this limitation, $\kappa = 1$ remains more flexible than the traditional mechanism ($\kappa = 0$).

2.3 User Subscription

Each MU is selfish, and will evaluate his/her payoff before subscribing to a data service $\mathcal{T} = \{Q, \Pi, \pi, \kappa\}$. The MU is willing to subscribe to the MNO's data service $\mathcal{T}$, if $\mathcal{T}$ brings him a non-negative payoff. The market share of the MNO under data plan $\mathcal{T}$ is

$$\Psi(\mathcal{T}) = \{(\theta, \beta) : \Theta(\mathcal{T}, \beta) \leq \theta \leq \theta_{\max}, 0 \leq \beta \leq 1\}. \tag{2.21}$$

2.3 User Subscription

For presentation convenience, we refer to $\Theta(\mathcal{T}, \beta)$ as the threshold valuation, i.e.,

$$\Theta(\mathcal{T}, \beta) \triangleq \left[\frac{\pi [\bar{d} - A_\kappa(Q)] - \Pi}{\beta A_\kappa(Q) - \bar{d}} + \pi \right] \frac{1}{\rho}. \tag{2.22}$$

Our analysis shows that the MU is more inclined to subscribe to MNO as their data valuation θ increases. However, the effect of network substitutability β is more nuanced. Depending on the given data valuation θ, the influence of network substitutability β can be categorized into three distinct scenarios.

- For Case 1 (i.e., $\pi[\bar{d} - A_\kappa(Q)] > \Pi$), when the MU's network substitutability is improving, the MU will subscribe to the MNO with a higher probability.
- For Case 2 (i.e., $\pi[\bar{d} - A_\kappa(Q)] < \Pi$), when the MU's network substitutability is improving, the MU will subscribe to the MNO with a lower probability.
- For Case 3 (i.e., $\pi[\bar{d} - A_\kappa(Q)] = \Pi$), the MU's network substitutability cannot change the subscription decision.

Figure 2.2 illustrates the three cases above. The gray region represents $\Psi(\mathcal{T})$. For the case shown in Fig. 2.2a, the per-unit fee π is large compared to the average

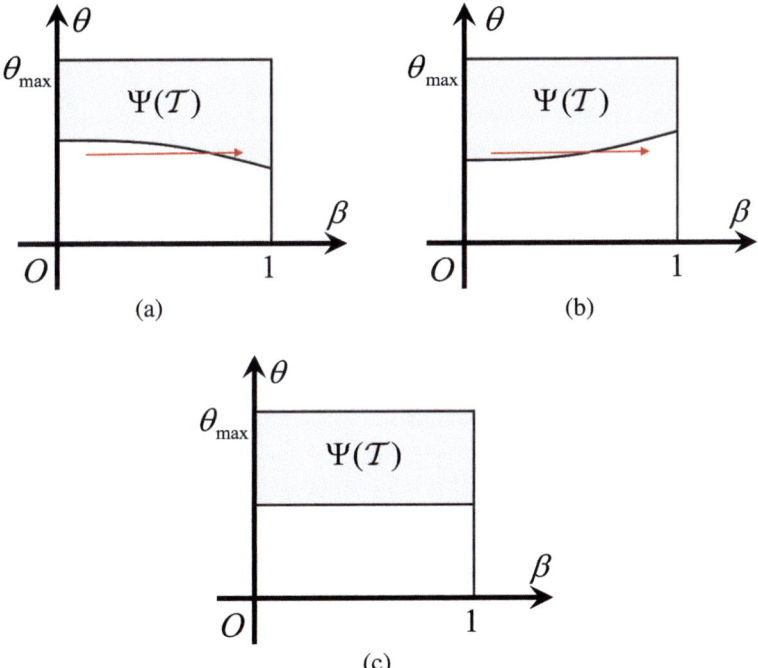

Fig. 2.2 Illustration of different market partitions. Gray region: the subscribers $\Psi(\mathcal{T})$. (**a**) Case 1. (**b**) Case 2. (**c**) Case 3

payment per unit data. Hence, the MUs with a better alternative network are more likely to become the MNO's subscribers. For the case shown in Fig. 2.2b, the per-unit fee π is small compared to the average payment per unit data. Hence, the MUs with a better alternative network are less likely to become the MNO's subscribers. These MUs are obviously not willing to pay for the subscription fee, given the small per-unit fee and good alternative networks. For the case shown in Fig. 2.2c, the average payment for unit data and the per-unit fee π just satisfy the condition $\pi[\bar{d} - A_\kappa(Q)] = \Pi$. In this case, the quality of alternative network cannot change the subscription decisions of MUs.

2.4 MNO's Data Service Optimization

Now we analyze how the MNO optimizes the data service $\mathcal{T} = \{Q, \Pi, \pi, \kappa\}$ to maximize its profit. Mathematically, the MNO's data service optimization problem is given by

$$\begin{aligned}
\{Q^*, \Pi^*, \pi^*, \kappa^*\} = \arg\max \quad & \tilde{W}(Q, \Pi, \pi, \kappa) \\
\text{s.t.} \quad & Q \geq 0, \\
& \Pi \geq 0, \\
& \pi \geq 0 \\
& \kappa \in \{0, 1, 2\},
\end{aligned} \quad (2.23)$$

where $\tilde{W}(Q, \Pi, \pi, \kappa)$ represents the MNO's expected profit under the market share presented in Sect. 2.3. In the subsequent, we introduce the MNO's optimal pricing strategy in Sect. 2.4.1, the optimal data cap in Sect. 2.4.2, and the optimal data mechanism in Sect. 2.4.3.

2.4.1 Optimal Pricing Strategy

We investigate the MNO's optimal pricing strategy given the data cap and the data mechanism. The results are presented in Theorem 2.1.

Theorem 2.1 *To maximize the MNO's profit, the subscription fee Π^* and the per-unit fee π^* should satisfy*

$$\begin{cases} \Pi^* = \pi^* \left[\bar{d} - A_\kappa(Q)\right], \\ H\left(\dfrac{\pi^*}{\rho}\right) + \dfrac{\pi^* - c}{\rho} \cdot h\left(\dfrac{\pi^*}{\rho}\right) = 1, \end{cases} \quad (2.24)$$

2.4 MNO's Data Service Optimization

where $h(\cdot)$ and $H(\cdot)$ are the probability density function and the cumulative distribution function of the data valuation θ. Furthermore, π^* is unique for any θ distribution satisfying increasing failure rate (IFR).[1]

The analytical results in Theorem 2.1 lead to several insightful observations, which are summarized as follows: First, π^* increases in ρ and c. It is not affected by the data mechanism κ or the data cap Q. Second, the optimal subscription fee $\Pi^*(Q, \kappa)$ increases in the data cap Q. Moreover, a better time flexibility corresponds to a higher subscription fee. That is, we have $\Pi^*(Q, 0) < \Pi^*(Q, 1) < \Pi^*(Q, 2)$. Third, under the optimal pricing strategy specified in Theorem 2.1, the MNO's market share is

$$\Psi\left(Q, \Pi^*, \pi^*, \kappa\right) = \left\{(\theta, \beta) : \frac{\pi^*}{\rho} \leq \theta \leq \theta_{\max}, 0 \leq \beta \leq 1\right\}. \tag{2.25}$$

This pricing approach demonstrates that the MNO primarily screens subscribers according to their data valuations, effectively disregarding differences in network substitutability. The economic rationale for this strategy lies in the dual revenue streams it generates: subscription fees from users with good network alternatives and overage payments from those with limited substitution options. Consequently, the MNO has no economic justification to exclude either user category from its service offerings.

2.4.2 Optimal Data Cap

Building upon the pricing strategy established in Theorem 2.1, we now examine the MNO's optimal data cap determination. For analytical clarity in deriving key insights, we adopt a linear capacity cost function of the form $J(Q) = z \cdot Q$, where z represents the marginal capacity cost parameter. The resulting optimal data cap characterization is presented in Theorem 2.2.

Theorem 2.2 *When the MNO adopts mechanism κ, the optimal data cap $Q^*(\kappa)$ is given by*

$$\left|A'_\kappa\left(Q^*(\kappa)\right)\right| = \Omega(\rho, c, z), \tag{2.26}$$

where $\Omega(\rho, c, z)$ is given by

$$\Omega(\rho, c, z) \triangleq \frac{z \cdot h\left(\frac{\pi^*}{\rho}\right)}{\bar{\beta}\rho\left[1 - H\left(\frac{\pi^*}{\rho}\right)\right]^2}. \tag{2.27}$$

---

[1] A distribution $h(\cdot)$ satisfies IFR condition if $h(\theta)/[1 - H(\theta)]$ is increasing in θ [47]. Many distributions (e.g., uniform distribution, gamma distribution, and normal distribution) satisfy this condition [48].

Theorem 2.2 establishes that the MNO's optimal data cap strategy remains consistent across all possible data mechanisms κ. Specifically, the profit-maximizing solution requires setting a data cap where the resulting marginal overage data consumption equals $\Omega(\rho, c, z)$. We therefore designate $\Omega(\rho, c, z)$ as the critical threshold for marginal overage consumption that must be attained for profit maximization. While intuition might suggest that enhanced time flexibility would allow for a reduced data cap (thereby lowering capacity costs through $J(Q) = z \cdot Q$), our analysis reveals an important counterintuitive finding: **Improved time flexibility does not invariably correspond to a smaller optimal data cap**. To elucidate this non-intuitive result, we analyze two distinct data mechanisms $i, j \in \{0, 1, 2\}$, where j provides superior time flexibility, i.e., $\mathcal{F}_i < \mathcal{F}_j$.

2.4.3 Optimal Data Mechanism

Now we are ready to present the MNO's profit-maximizing data mechanism κ^*. Based on the results in Sects. 2.4.1 and 2.4.2, we find that a data mechanism offering a better time flexibility leads to more profit for the MNO. That is, the rollover data mechanism of China Mobile leads to a larger profit.

2.5 MNO's Price Discrimination

We introduce how the MNO leverages price discrimination to enlarge its profit earned from mobile data services. To this end, the MNO will proceed with the contract design (with multiple data caps) to tackle the MUs' private information.

2.5.1 Contract Formulation

We consider K different data valuation, denoted by the set $\Theta = \{\theta_k : 1 \le k \le K\}$. Moreover, we consider M network substitutability types, denoted by the set $\mathcal{B} = \{\beta_m : 1 \le m \le M\}$. Moreover, MUs' types are indexed in the ascending order, i.e., $\theta_1 < \theta_2 < \ldots < \theta_K$ and $\beta_1 < \beta_2 < \ldots < \beta_M$. To sum up, we have a total of KM types of MUs in the telecommunication market. We use $q_{m,k}$ to denote the fraction of type-(β_m, θ_k) MU. The revelation principle [49] indicates it enough for the MNO to consider a class of contract items that enables MUs to truthfully reveal their private information (i.e., MU type). This means that the MNO only needs to design a contract with KM items, one for each MU type. We let

$$\Phi(\mathcal{B}, \Theta) = \{\phi_{m,k} : 1 \le m \le M, 1 \le k \le K\} \tag{2.28}$$

2.5 MNO's Price Discrimination

denote the contract, where $\phi_{m,k} = \{Q_{m,k}, \Pi_{m,k}\}$ is the contract item designed for the type-(β_m, θ_k) MUs. In general, we say that a contract is *feasible* if and only if it induces each MU to select the contract item designed for this type. It is obvious that a contract is *feasible* if and only if it satisfies two fundamental economic constraints: Individual Rationality (IR) and Incentive Compatibility (IC).

Definition 2.2 (Individual Rationality) A contract satisfies the Individual Rationality (IR) condition if each type-(β_m, θ_k) MU can obtain a non-negative payoff by selecting the contract item $\phi_{m,k}$ designed for this MU type, i.e.,

$$S(\phi_{m,k}, \beta_m, \theta_k) \geq 0. \tag{2.29}$$

We let $\phi_{m,k} \succcurlyeq 0$ denote the IR condition for the contract item $\phi_{m,k}$.

Definition 2.3 (Incentive Compatibility) A contract satisfies the Incentive Compatibility (IC) condition if MUs of type-(β_m, θ_k) can maximize its payoff by subscribing to the contract item $\phi_{m,k}$, i.e.,

$$S(\phi_{m,k}, \beta_m, \theta_k) \geq S(\phi_{n,l}, \beta_m, \theta_k), \ \forall \ (n,l) \neq (m,k). \tag{2.30}$$

Our subsequent analysis relies on the notion of Pairwise Incentive Compatibility (PIC). Essentially, the original $KM(KM-1)$ incentive compatibility (IC) constraints can be reduced to $KM(KM-1)/2$ PIC conditions, each corresponding to a unique pair of MUs.

Definition 2.4 (Pairwise Incentive Compatibility) Two contract items $\phi_{m,k}$ and $\phi_{n,l}$ satisfy the pairwise IC condition if and only if

$$\begin{cases} S(\phi_{m,k}, \beta_m, \theta_k) \geq S(\phi_{n,l}, \beta_m, \theta_k), \\ S(\phi_{n,l}, \beta_n, \theta_l) \geq S(\phi_{m,k}, \beta_n, \theta_l). \end{cases} \tag{2.31}$$

We use the following notation to represent the PIC condition for the contract items $\phi_{m,k}$ and $\phi_{n,l}$

$$\phi_{m,k} \stackrel{IC}{\Longleftrightarrow} \phi_{n,l}. \tag{2.32}$$

Recall that the MNO's revenue from an MU consists of the subscription fee and the overage fee. The MNO's expected revenue $R(\Phi)$ under a *feasible* contract Φ is given by

$$R(\Phi) = \sum_{k=1}^{K} \sum_{m=1}^{M} q_{m,k} \big[\underbrace{\Pi_{m,k}}_{\text{subscription}} + \underbrace{P(Q_{m,k}, \beta_m)}_{\text{overage}} \big]. \tag{2.33}$$

The MNO incurs two fundamental types of costs: the capacity cost and operational cost. Arising from infrastructure investments to support network capacity, the MNO's capital expenditure is significantly influenced by its data cap policy. Empirical evidence suggests that data caps serve as an effective tool for congestion management and capacity allocation [50]. We model the capacity cost for a type-(β_m, θ_k) subscriber as a monotonically increasing function $J(Q)$ of the allocated data cap Q [38]. This formulation captures the intuitive relationship whereby higher data caps lead to increased network congestion and greater required infrastructure investments. Related to ongoing system management and maintenance [51], MNO's operational cost varies with subscribers' actual data usage patterns. After the MNO decides which data plan to implement, the subscribers' total data consumption will influence the MNO's operational expense. Therefore, the MNO's operational cost caused by a type-(β_m, θ_k) subscriber with data cap Q can be formulated as $c \cdot U(Q, \beta_m)$, where c is the MNO's marginal cost for the system management [50], and $U(Q, \beta_m) = \bar{d} - \beta_m A(Q)$ is the type-(β_m, θ_k) MU's expected data consumption. Overall, the MNO's expected cost $C(\Phi)$ under a *feasible* contract Φ is

$$C(\Phi) = \sum_{k=1}^{K} \sum_{m=1}^{M} \big[\underbrace{c \cdot U(Q_{m,k}, \beta_m)}_{\text{Operational cost}} + \underbrace{J(Q_{m,k})}_{\text{Capacity cost}} \big] q_{m,k}. \quad (2.34)$$

Accordingly, the MNO's expected profit under a *feasible* contract Φ is the difference between its revenue and cost, given by

$$W(\Phi) = R(\Phi) - C(\Phi). \quad (2.35)$$

To achieve the goal of price discrimination, the MNO is faced with the following optimal contract design problem:

Problem 2.1 (Optimal Contract Design)

$$\max_{\Phi} W(\Phi) \quad (2.36)$$
$$s.t. \ (2.29), (2.30).$$

Note that Problem 2.1 requires the MNO to induce the IC and IR conditions, so that each MU will participate and truthfully reveals the private type by selecting the contract item designed for this MU type. However, Problem 2.1 involves multi-dimensional MU types, thus is challenging in general. When contract problems involve one-dimensional MU types, the single-crossing condition ensures that the allocation rule remains monotone. As a result, the methodology adopted in prior studies [52–54] can effectively eliminate many non-binding IC and IR constraints, simplifying the contract problem. However, this approach does not readily extend

2.5 MNO's Price Discrimination

to cases with two-dimensional MU types. To overcome this limitation, we propose incorporating the marginal rate of substitution.

2.5.2 Marginal Rate of Substitution

In economic theory, an indifference curve represents combinations of goods that provide a consumer with equal utility. Applying this concept to our model, we can depict the MU's indifference curve on the contract plane, where the axes correspond to the data allowance (Q) and the subscription cost (Π), as illustrated in Fig. 2.3. On the plane of (Q, Π), the following indifference curve ensures that the type-(β, θ) MU gains a fixed payoff S

$$S = \theta[\bar{d} - \beta A(Q)] - \pi(1-\beta)A(Q) - \Pi. \tag{2.37}$$

As shown in Fig. 2.3, the indifference curve is increasing and concave in the data cap Q. As an MU's indifference curve shifts downward, his payoff increases because of the decreasing subscription fee.

The marginal rate of substitution (MRS), defined as the slope of an indifference curve, measures the maximum amount of one good a consumer is willing to sacrifice for an additional unit of another good without altering their overall utility. In our problem, (2.37) could be expressed as follows:

$$\Pi(\theta, \beta, Q, \pi) = -S + \theta\left[\bar{d} - \beta A(Q)\right] - \pi(1-\beta)A(Q). \tag{2.38}$$

The MRS of a type-(β, θ) MU is given by

$$\sigma(Q, \beta, \theta) \triangleq \frac{\partial \Pi(\theta, \beta, Q, \pi)}{\partial Q} \\ = -[\theta\beta + \pi(1-\beta)] \cdot \frac{\partial A(Q)}{\partial Q}. \tag{2.39}$$

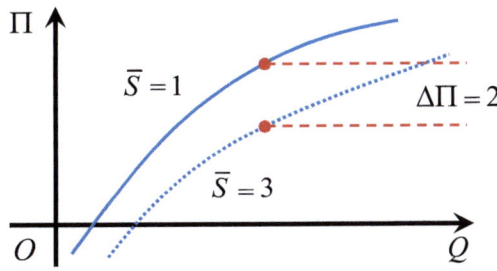

Fig. 2.3 Illustration of an MU's indifference curves

Note that MRS depends on both the MU's private information (β, θ) and the data cap Q. Specifically, $\sigma(Q, \beta, \theta)$ quantifies a type-(β, θ) MU's willingness-to-pay for an incremental unit of data under a given data cap Q. We sort and index the KM MU types according to the willingness-to-pay in an ascending order, i.e.,

$$\Lambda_1(Q), \Lambda_2(Q), \ldots, \Lambda_{KM}(Q), \qquad (2.40)$$

In this case, we have

$$\sigma(Q, \Lambda_1) \leq \sigma(Q, \Lambda_2) \leq \ldots \leq \sigma(Q, \Lambda_{KM}). \qquad (2.41)$$

Note that the revised MU ordering in Eq. (2.40) is independent of the data cap. Specifically, for any two distinct data caps $Q \neq Q'$, the following holds:

$$\Lambda_i(Q) = \Lambda_i(Q'), \ \forall\ 1 \leq i \leq KM. \qquad (2.42)$$

That is, the MU ordering in (2.40) remains invariant despite variations in $\sigma(Q, \Lambda_i)$ with respect to the data cap Q. This invariance stems from the separable structure of an MU's willingness-to-pay $\sigma(Q, \Lambda_i)$ in (2.39), which decomposes into distinct components for MU types (θ, β) and the data cap Q. For notational simplicity, we will henceforth use Λ_i to represent an MU type under the ordering defined in (2.41), and denote by $\phi_i = Q_i, \Pi_i$ the contract item designed for type-Λ_i MUs. To facilitate our later discussion, we summarize the mapping from (β_m, θ_k) to Λ_1 and Λ_{KM} as

$$\begin{cases} \Lambda_1 = \{\beta_M, \theta_1\}, \Lambda_{KM} = \{\beta_1, \theta_K\}, & \text{if } \theta_1 < \theta_K < \pi, \\ \Lambda_1 = \{\beta_M, \theta_1\}, \Lambda_{KM} = \{\beta_M, \theta_K\}, & \text{if } \theta_1 < \pi < \theta_K, \\ \Lambda_1 = \{\beta_1, \theta_1\}, \Lambda_{KM} = \{\beta_M, \theta_K\}, & \text{if } \pi < \theta_1 < \theta_K. \end{cases} \qquad (2.43)$$

Furthermore, we define the smallest-payoff MU type $\Lambda_\epsilon(Q, \Pi)$ under the contract item (Q, Π) as

$$\Lambda_\epsilon(Q, \Pi) \triangleq \arg\min_{\Lambda_i} S(Q, \Pi, \Lambda_i), \qquad (2.44)$$

and specify the type-Λ_ϵ MU as follows:

$$\begin{cases} \Lambda_\epsilon = \{\beta_1, \theta_1\}, & \text{if } \theta_1 < \theta_K < \pi, \\ \Lambda_\epsilon = \{\beta_1, \theta_1\}, & \text{if } \theta_1 < \pi < \theta_K, \\ \Lambda_\epsilon = \{\beta_M, \theta_1\}, & \text{if } \pi < \theta_1 < \theta_K. \end{cases} \qquad (2.45)$$

2.5.3 Necessary Conditions for Feasible Contract

Now we leverage MUs' *willingness-to-pay* to study the necessary conditions for a contract to be feasible. Lemmas 2.2 and 2.3 present the main results.

Lemma 2.2 *For any feasible contract* $\Phi(\mathcal{B}, \Theta)$, $Q_i < Q_j$ *if and only if* $\Pi_i < \Pi_j$.

Lemma 2.2 demonstrates that, within the feasible contract set, an increase in the data cap necessitates a higher subscription fee.

Lemma 2.3 *For any feasible contract* $\Phi(\mathcal{B}, \Theta)$, *if* $\sigma(Q, \Lambda_i) > \sigma(Q, \Lambda_j)$ *for all* Q, *then* $Q_i \geq Q_j$.

Lemma 2.3 indicates that the optimal contract assigns larger data caps to MUs with greater willingness-to-pay. We leverage Fig. 2.4 to illustrate the key insights of Lemma 2.3.

Consider a type-Λ_j MU for whom the contract item ϕ_j is designed, represented by the red dot in Fig. 2.4. The red indifference curve l_j corresponds to this MU's utility when selecting ϕ_j. For a type-Λ_i MU, the blue indifference curve l_i reflects the same utility level obtained by choosing ϕ_j (despite ϕ_j not being tailored for this type). Note that the curve l_i is steeper than the curve l_j. Mathematically, the inequality $\sigma(Q, \Lambda_i) > \sigma(Q, \Lambda_j)$ holds for all $Q > 0$. This implies that type-Λ_i MUs exhibit strictly greater willingness-to-pay than type-Λ_j MUs, regardless of the data cap value. Furthermore, a downward shift in an MU's indifference curve signifies higher utility, attributable to the reduced subscription fee at any given data cap. To ensure the PIC condition $\phi_i \overset{IC}{\Longleftrightarrow} \phi_j$, the contract item ϕ_i (intended for the type-Λ_i MUs) must locate

- below or on the blue square curve l_i (ensuring Λ_i prefers ϕ_i over ϕ_j), and
- above or on the red circle curve l_j (ensuring Λ_j prefers ϕ_j over ϕ_i).

Geometrically, ϕ_i must reside within the blue shaded region of Fig. 2.4. We prove this result by contradiction. Assume the contrary, then at least one of the following scenarios must hold:

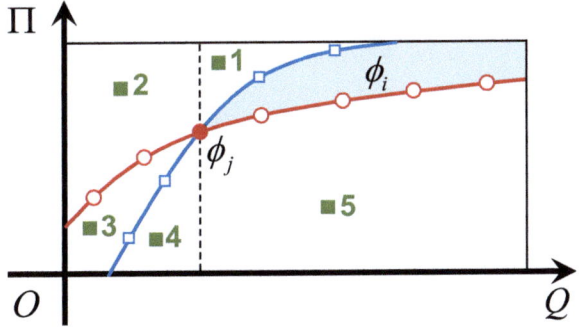

Fig. 2.4 An illustration of the feasible region of contract item ϕ_i

- **Scenario 1:** The contract item ϕ_i (represented by the green squares labeled 1–3 in Fig. 2.4) lies above the type-Λ_i indifference curve l_i. This configuration requires the indifference curve l_i to shift upward (resulting in decreased payoff) to intersect any of the green-square contracts. However, the type-Λ_i MU can obtain strictly higher payoff by selecting the red-dot contract ϕ_j rather than choosing ϕ_i, thereby violating the Pairwise Incentive Compatibility (PIC) condition for type-Λ_i MUs.
- **Scenario 2:** The contract item ϕ_i (represented by green squares 4 and 5 in Fig. 2.4) lies below the type-Λ_j indifference curve l_j. This configuration requires curve l_j to shift downward (indicating increased payoff) to intersect with any green-square contract. Consequently, type-Λ_j MUs obtain strictly higher payoff by selecting ϕ_i rather than their designated contract, violating their Pairwise Incentive Compatibility (PIC) condition.

The preceding analysis demonstrates that the contract item ϕ_i must lie within the blue region of Fig. 2.4. This geometric constraint necessarily implies the monotonicity condition $Q_i \geq Q_j$ for the optimal contract design. Based on Lemmas 2.2 and 2.3, we derive the following necessary conditions for a feasible contract:

Theorem 2.3 (Necessary Conditions for Feasibility) *The feasible contract satisfies the following conditions*

$$\begin{cases} Q_1 \leq Q_2 \leq \ldots \leq Q_{KM}, \\ \Pi_1 \leq \Pi_2 \leq \ldots \leq \Pi_{KM}. \end{cases} \quad (2.46)$$

2.5.4 Sufficient Conditions for Feasible Contract

Before analyzing the sufficient conditions of feasible contract, we first introduce two transitivity properties.

Lemma 2.4 (PIC-Transitivity) *For any $i_1 < i_2 < i_3$, we have*

$$\text{if } \phi_{i_1} \overset{IC}{\Longleftrightarrow} \phi_{i_2} \text{ and } \phi_{i_2} \overset{IC}{\Longleftrightarrow} \phi_{i_3}, \text{ then } \phi_{i_1} \overset{IC}{\Longleftrightarrow} \phi_{i_3}. \quad (2.47)$$

The transitivity property of PIC simplifies the contract problem by reducing the number of required conditions. Instead of verifying all $KM(KM-1)/2$ PIC constraints, it suffices to check only $KM - 1$ conditions—specifically, those between neighboring MU types, i.e., $\phi_i \overset{IC}{\Longleftrightarrow} \phi_{i+1}$ for $i \in \{1, 2, \ldots, KM - 1\}$.

Lemma 2.5 (IR-Transitivity) *Suppose all the PIC conditions hold, then we have*

$$\text{if } \phi_\epsilon \succcurlyeq 0, \text{ then } \phi_i \succcurlyeq 0, \ \forall\, i \neq \epsilon.$$

2.5 MNO's Price Discrimination

Recall that type-Λ_ϵ MUs (defined in (2.44)) receive the lowest payoff under any contract item. By Lemma 2.5, ensuring all PIC conditions means we only need to verify the IR constraint for type-Λ_ϵ. Therefore, the KM original IR conditions collapse to a single condition: $\phi_\epsilon \succcurlyeq 0$.

Before deriving sufficient conditions, we introduce a new concept for each MU, which is called the virtual payoff increment in this book. This concept is related to the contract items designed for its neighbor type. Mathematically, the virtual payoff increment for type-Λ_i is

$$\eta^-(\Lambda_i, Q_i, Q_{i-1}) = L(Q_i, \Lambda_i) - L(Q_{i-1}, \Lambda_i), \tag{2.48a}$$

$$\eta^+(\Lambda_i, Q_i, Q_{i+1}) = L(Q_i, \Lambda_i) - L(Q_{i+1}, \Lambda_i). \tag{2.48b}$$

Leveraging the results from Lemmas 2.2 to 2.5, we derive the following sufficient conditions to ensure contract feasibility:

Theorem 2.4 (Sufficient Conditions) *The contract $\Phi(\mathcal{B}, \Theta)$ is feasible if the following conditions hold.*

(1) $Q_1 \leq Q_2 \leq \ldots \leq Q_{KM}$,
(2) for $i = \epsilon$,

$$\Pi_\epsilon \leq L(Q_\epsilon, \Lambda_\epsilon), \tag{2.49}$$

(3) for all $i = 1, 2, \ldots, \epsilon - 1$,

$$\Pi_i \leq \Pi_{i+1} + \eta^+(\Lambda_i, Q_i, Q_{i+1}), \tag{2.50a}$$

$$\Pi_i \geq \Pi_{i+1} - \eta^-(\Lambda_{i+1}, Q_{i+1}, Q_i). \tag{2.50b}$$

(4) for all $i = \epsilon + 1, \epsilon + 2, \ldots, KM$,

$$\Pi_i \leq \Pi_{i-1} + \eta^-(\Lambda_i, Q_i, Q_{i-1}), \tag{2.51a}$$

$$\Pi_i \geq \Pi_{i-1} - \eta^+(\Lambda_{i-1}, Q_{i-1}, Q_i), \tag{2.51b}$$

We now examine the intuition behind Theorem 2.4. *Condition 1)* satisfies the necessary requirements from Theorem 2.3. *Condition 2)* ensures IR compliance for type-Λ_ϵ MUs ($\phi_\epsilon \succcurlyeq 0$), which by Lemma 2.5 extends to all MU types. *Condition 3)* and *Condition 4)* enforce PIC between adjacent MU types ($\phi_i \xleftrightarrow{\text{IC}} \phi_{i+1}$ for $1 \leq i \leq KM - 1$), sufficient for global IC via Lemma 2.4. Specifically, the inequality (2.50a) ensures type-Λ_i MUs not willing to select the contract item ϕ_{i+1}, i.e., $S(\phi_i, \Lambda_i) \geq S(\phi_{i+1}, \Lambda_i)$; the inequality (2.50b) ensures type-Λ_{i+1} MUs not willing to select the contract item ϕ_i, i.e., $S(\phi_{i+1}, \Lambda_{i+1}) \geq S(\phi_i, \Lambda_{i+1})$. Furthermore, (2.51) exhibits the similar intuitions.

2.5.5 Optimal Contract Design

We analyze the MNO's optimal contract design problem using the established necessary and sufficient conditions. Our analysis proceeds in two key steps.

Step 1 We derive the MNO's optimal prices $\{\Pi_i^*(\boldsymbol{Q}), 1 \leq i \leq KM\}$ given a *feasible* choice of data caps $\boldsymbol{Q} = \{Q_i, 1 \leq i \leq KM\}$ where $Q_1 \leq Q_2 \leq \ldots \leq Q_{KM}$. This corresponds to Problem 2.2.

Problem 2.2 (Optimal Prices)

$$\max \sum_{i=1}^{KM} \Big[\Pi_i + P(Q_i, \Lambda_i) - c \cdot U(Q_i, \Lambda_i) - J(Q_i)\Big] q(\Lambda_i) \tag{2.52}$$

s.t. (2.49), (2.50), (2.51)

var. $\{\Pi_i : \forall 1 \leq i \leq KM\}$.

Note that the constraints (2.49), (2.51), and (2.50) correspond to sufficient conditions in Theorem 2.4. In this case, the obtained solution $\{\Pi_i^*(\boldsymbol{Q}), 1 \leq i \leq KM\}$ as well as $\boldsymbol{Q}$ could form a feasible contract. We characterize the optimal prices $\{\Pi_i^*(\boldsymbol{Q}), 1 \leq i \leq KM\}$ in Theorem 2.5.

Theorem 2.5 *Suppose that the data caps $\boldsymbol{Q}$ satisfy $Q_1 \leq Q_2 \leq \ldots \leq Q_{KM}$, then the MNO's optimal pricing policy is*

$$\Pi_i^*(\boldsymbol{Q}) = L(Q_i, \Lambda_i), \qquad \text{if } i = \epsilon, \tag{2.53a}$$

$$\Pi_i^*(\boldsymbol{Q}) = \Pi_{i+1}^*(\boldsymbol{Q}) + \eta^+(\Lambda_i, Q_i, Q_{i+1}), \qquad \text{if } i < \epsilon, \tag{2.53b}$$

$$\Pi_i^*(\boldsymbol{Q}) = \Pi_{i-1}^*(\boldsymbol{Q}) + \eta^-(\Lambda_i, Q_i, Q_{i-1}), \qquad \text{if } i > \epsilon. \tag{2.53c}$$

A comparison between Theorems 2.4 and 2.5 reveals that the MNO must set prices at the maximum level permitted by the IC and IR constraints.

Step 2 We substitute the obtained prices $\{\Pi_i^*(\boldsymbol{Q}), 1 \leq i \leq KM\}$ to the profit function of MNO, and then derive the optimal data cap $\boldsymbol{Q}^* = \{Q_i^*, 1 \leq i \leq KM\}$. This corresponds to Problem 2.3. To enhance readability, we define the virtual payoff difference for a given Q as the difference in payoffs between a type-Λ_i MU and its adjacent types, Λ_{i-1} and Λ_{i+1}:

$$\rho_i^-(Q) \triangleq L(Q, \Lambda_i) - L(Q, \Lambda_{i-1}), \tag{2.54a}$$

$$\rho_i^+(Q) \triangleq L(Q, \Lambda_i) - L(Q, \Lambda_{i+1}). \tag{2.54b}$$

2.5 MNO's Price Discrimination

Substituting (2.53) into Problem 2.2's objective and leveraging (2.54), we express the MNO's total profit as:

$$\sum_{i=1}^{KM} G_i(Q_i), \qquad (2.55)$$

where $G_i(\cdot)$ is given by

$$G_i(Q) = \begin{cases} q(\Lambda_i)V(Q, \Lambda_i) - q(\Lambda_i)[c \cdot U(Q, \Lambda_i) + J(Q)], \\ \qquad\qquad \text{if } i \in \{1, KM\}, \\ q(\Lambda_i)V(Q, \Lambda_i) + h^i \rho_i^-(Q) - q(\Lambda_i)[cU(Q, \Lambda_i) + J(Q)], \\ \qquad\qquad \text{if } i \in \{2, 3, \dots, \epsilon - 1\}, \\ q(\Lambda_i)V(Q, \Lambda_i) + h^i \rho_i^-(Q) + h_i \rho_i^+(Q) \\ \qquad - q(\Lambda_i)[cU(Q, \Lambda_i) + J(Q)], \text{if } i = \epsilon, \\ q(\Lambda_i)V(Q, \Lambda_i) + h_i \rho_i^+(Q) - q(\Lambda_i)[cU(Q, \Lambda_i) + J(Q)], \\ \qquad\qquad \text{if } i \in \{\epsilon+1, \epsilon+2, \dots, KM-1\}, \end{cases} \qquad (2.56)$$

Moreover, h^i and h_i are defined as follows:

$$h^i = \sum_{t=1}^{i-1} q(\Lambda_t),$$
$$h_i = \sum_{t=i+1}^{KM} q(\Lambda_t), \qquad (2.57)$$

which are two constants related to the distribution of the MU types. Now we obtain Problem 2.3 as follows:

Problem 2.3 (Optimal Data Caps)

$$\max \sum_{i=1}^{KM} G_i(Q_i) \qquad (2.58a)$$

$$\text{s.t. } Q_1 \leq Q_2 \leq \dots \leq Q_{KM} \leq D \qquad (2.58b)$$

$$Q_i \in \mathbb{N}, \; \forall i \in \{1, 2, \dots, KM\} \qquad (2.58c)$$

$$\text{var. } Q_i, 1 \leq i \leq KM. \qquad (2.58d)$$

The nonlinear integer programming formulation in Problem 2.3 possesses two key characteristics: (1) a separable objective function in terms of each Q_i, and (2) monotonic decision variables. Furthermore, the problem's convexity is contingent upon both the full spectrum of MU types Λ_i (where $1 \leq i \leq KM$) and their respective probability distributions $q(\Lambda_i)$.

Prior studies (e.g., [52–54]) have typically addressed Problem 2.3 using monotonicity relaxation. This approach involves two key steps:

- Relaxation: The monotonicity constraints (2.58b) are temporarily ignored, allowing each $G_i(\cdot)$ to be maximized independently over its corresponding Q_i.
- Adjustment: If the relaxed solution violates (2.58b), it is iteratively modified using the algorithm from [52] to restore feasibility.

While computationally efficient—owing to its decomposition into single-variable optimizations—this method has two limitations:

- Local Optimality: In non-convex cases, the adjusted solution is only guaranteed to be locally optimal [55].
- Sub-Optimality Gap: The deviation from the global optimum lacks a tractable analytical characterization.

To overcome these issues, Sect. 2.5.6 introduces the Dynamic Quota Allocation Algorithm, which efficiently guarantees global optimality.

2.5.6 Dynamic Quota Allocation (DQA) Algorithm

The DQA Algorithm is rooted in dynamic programming principles, which involve solving complex problems by recursively dividing them into smaller, more manageable sub-problems [56]. To apply this approach, we decompose Problem 2.3 by leveraging the separable structure of the objective function (2.58a) and the monotonicity conditions (2.58b). The following section outlines the formulation of these sub-problems.

Level-(n, q) Subproblem We define Problem 2.4 as the level-(n, q) sub-problem of the original Problem 2.3. This sub-problem determines the optimal data caps for the first n MU types (i.e., type-1 to type-n, where $1 \leq n \leq KM$), subject to an upper bound q ($0 \leq q \leq D$) on the cap size. Here, KM represents the total number of MU types, and D denotes the maximum monthly data demand among all MUs. Notably, when $n = KM$ and $q = D$, the level-(KM, D) sub-problem reduces to Problem 2.3, as caps exceeding D are unnecessary.

Problem 2.4 For $1 \leq n \leq KM$ and $0 \leq q \leq D$, we obtain the following level-(n, q) sub-problem

2.5 MNO's Price Discrimination

$$H(n,q) \triangleq \arg\max \sum_{i=1}^{n} G_i(Q_i) \tag{2.59a}$$

$$\text{s.t. } Q_1 \leq Q_2 \leq \ldots \leq Q_n \leq q \tag{2.59b}$$

$$Q_i \in \mathbb{N}, \ \forall i \in \{1, 2, \ldots, n\} \tag{2.59c}$$

$$\text{var: } Q_i, 1 \leq i \leq n. \tag{2.59d}$$

Let $H(n,q)$ denote the optimal value and $\boldsymbol{Q}^{\star}(n,q) = \{Q_i^{\star}(n,q) \mid 1 \leq i \leq n\}$ denote the optimal solution to the level-(n,q) sub-problem (2.59). Given that the level-(KM, D) sub-problem is equivalent to Problem 2.3, it follows that:

- The optimal value of Problem 2.3 is $H(KM, D)$.
- The optimal data caps in Problem 2.3 is given by $\boldsymbol{Q}^{\star}(KM, D)$, i.e., $Q_i^* = Q_i^{\star}(KM, D)$ for all $1 \leq i \leq KM$.

Next, we demonstrate that knowing $H(n,q)$ for all $1 \leq n \leq KM$ and $0 \leq q \leq D$ enables direct computation of $\boldsymbol{Q}^{\star}(KM, D)$. To establish this relationship, we first analyze key properties of $H(n,q)$ in Propositions 2.1 and 2.2.

Proposition 2.1 *For any $n \in \{2, 3, \ldots, KM\}$ and $q \in \{0, 1, \ldots, D\}$, the following recursive relation holds*

$$H(n,q) = \max_{x \in \mathbb{N}} H(n-1, x) + G_n(x) \tag{2.60a}$$

$$\text{s.t. } x \leq q. \tag{2.60b}$$

Proposition 2.2 *For any $n \in \{1, 2, \ldots, KM\}$, the function $H(n,q)$ is positively related to the data cap q. Moreover, there always exists a critical point $\hat{q}_n$ such that $H(n,q)$ remains the same for any $q \geq \hat{q}_n$.*

Leveraging the recursive structure established in Proposition 2.1 and the critical points $\{\hat{q}_i \mid 1 \leq i \leq KM\}$ identified in Proposition 2.2, we can derive the optimal solution to the level-(KM, D) sub-problem (equivalent to Problem 2.3) through Theorem 2.6.

Theorem 2.6 *The optimal solution of the level-(KM, D) sub-problem is*

$$Q_i^{\star}(KM, D) = \begin{cases} \hat{q}_i, & \text{if } i = KM, \\ \min\{\hat{q}_i, Q_{i+1}^{\star}(KM, D)\}, & \text{if } i < KM. \end{cases} \tag{2.61}$$

The key remaining challenge lies in efficiently computing the $H(n,q)$ table. We address this through our proposed DQA Algorithm, which leverages the recursive properties established in Proposition 2.1. The complete computational procedure is presented in Algorithm 1. The input of Algorithm 1 includes the MU types and the distribution. The output of Algorithm 1 is the table of $\{H(n,q) : 1 \leq n \leq$

Algorithm 1: Dynamic Quota Allocation (DQA)

Input : All MU types Λ_i and the distribution $q(\Lambda_i)$.
Output: $H(n, q)$ for all $1 \leq n \leq KM$ and $0 \leq q \leq D$.
1 Initial $H(n, q) = 0$, $\forall\, 1 \leq n \leq KM$ and $0 \leq q \leq D$.
2 **for** $n = 1$ **to** KM **do**
3   **for** $q = 0$ **to** D **do**
4   **if** $n = 1$ **then**
5           $H(n, q) := \max_{x \leq q} G_n(x)$.
6   **else**
7           $H(n, q) := \arg\max_{x \leq q} H(n-1, x) + G_n(x)$.

$KM, 0 \leq q \leq D\}$. To obtain this table, we compute $H(1, q)$ in Line 5, and then compute $H(n, q)$ for all $n \geq 2$ in Line 7, which utilizes the recursive property in Proposition 2.1.

2.6 Summary

This chapter analyzes how rollover data services influence pricing and plan design in monopolistic telecommunications markets. Specifically, we examine the pricing strategy optimization by evaluating MNO's optimal pricing under various rollover mechanisms and designing multi-tiered data plans with time-flexible allocation. Moreover, we investigate asymmetric information framework by modeling MUs with two-dimensional private characteristics (i.e., individual data valuation and network substitutability preferences). We propose a contract-theoretic approach by formulating the MNO's cap design as a multidimensional incentive-compatible contract problem. Moreover, we derive the optimal contracts across different rollover schemes. We find that the willingness-to-pay is quantified by the slope of MUs' indifference curves in contract space. The incentive-compatible allocations prioritize higher data caps for MUs with greater payment propensity. Finally, we develop a computationally efficient method to determine optimal contract terms.

Our analysis reveals that time-flexible rollover data services create a Pareto-improving outcome, benefiting both market participants:

- The MNO experiences increased profitability.
- The MUs achieve higher utility payoffs.

This mutually advantageous scenario provides critical theoretical grounding for examining competitive market dynamics in Chap. 3, where we extend this framework to the competitive settings.

Chapter 3
Time Flexibility in Competitive Market

Abstract Having established the optimal strategies for monopolistic markets, we now examine how competition between multiple MNOs affects the adoption and effectiveness of time-flexible data services. We will analyze the MNOs' pricing equilibrium and the data service adoption equilibrium. This chapter will answer whether all the MNOs should enhance the time flexibility by adopting the rollover data service.

Keywords Rollover data service · Time flexibility · Market competition

3.1 Market Model

We analyze a duopoly market where two heterogeneous MNOs compete for a shared pool of mobile users (MUs). Our model captures the MNO-level heterogeneity in Quality-of-Service (QoS) and operational costs, as well as the MU-level heterogeneity in data valuations. Each mobile network operator MNO-n ($n \in \{1, 2\}$) offers a data plan characterized by the 4-tuple:

- Data Allowance: Q_n represents the monthly data cap (in GB).
- Subscription Fee: Π_n denotes the fixed monthly charge (in $).
- Overage Pricing: π_n specifies the marginal cost per additional unit of data beyond Q_n (in $/GB).
- Data Mechanism: κ_n denotes the data mechanism that the MNO-n adopts. Specifically, $\kappa_n = \text{T}$ and $\kappa_n = \text{R}$ represent the traditional mechanism and the rollover mechanism, respectively. For $\kappa_n = \text{R}$, the rollover data "inherited" from the previous month is consumed prior to the current monthly data cap and expires at the end of the current month.

For notation consistency, we follow Chap. 2 and denote τ as an MU's rollover data obtained from the previous billing cycle, and let $Q^e_{\kappa_n}(\tau)$ denote the *effective cap* of the current month. To facilitate our later analysis, here we define V_{κ_n} as

$$V_{\kappa_n} \triangleq \bar{d} - \beta A_{\kappa_n}(Q_n), \tag{3.1}$$

which represents the MU's expected monthly data consumption under the data mechanism κ_n. In general, rollover mechanism R provides time flexibility that encourages MUs to consume more data. Hence we have

$$V_R > V_T. \tag{3.2}$$

Accordingly, the expected monthly payoff of *a type-θ $\mathcal{T}_n$ MU* is

$$\bar{U}_n(\mathcal{T}_n, \theta) = \rho_n V_{\kappa_n} \theta - \pi_n \left(\beta^{-1} - 1\right) (\bar{d} - V_{\kappa_n}) - \Pi_n, \tag{3.3}$$

where $\rho_n V_{\kappa_n}$ represents the MU's utility increment for unit data valuation increment under the subscription of MNO-n. In subsequent analysis, we adopt the payoff expression from (3.3). To explicitly highlight its dependence on both the data mechanism κ_n and pricing strategy $s_n = \Pi_n, \pi_n$, we may equivalently express the payoff as $\bar{U}_n(\kappa_n, s_n, \theta)$.

We analyze the strategic interaction between two MNOs competing through rollover mechanism selection and pricing strategies under symmetric data cap conditions (e.g., $Q_1 = Q_2 = $ 1GB). The duopoly competition unfolds as a three-stage sequential game:

- Stage I: MNOs simultaneously announce their data mechanisms κ_1 and κ_2.
- Stage II: MNOs determine the prices $s_1 = \{\Pi_1, \pi_1\}$ and $s_2 = \{\Pi_2, \pi_2\}$.
- Stage III: MUs make subscription choices based on the offered mechanisms, price vectors, and individual valuation parameters.

3.2 User Subscription

Before analyzing the duopoly market partition, we first introduce two critical concepts, i.e., threshold MU type and neutral MU type.

Definition 3.1 The MNO-n's threshold MU type, denoted by $\sigma_n \in [0, \theta_{\max}]$, corresponds to the MU type who obtains a zero expected payoff (i.e., $\bar{U}_n(\kappa_n, s_n, \sigma_n) = 0$).

$$\sigma_n(\kappa_n, s_n) \triangleq \frac{\Pi_n + \pi_n(\beta^{-1} - 1)(\bar{d} - V_{\kappa_n})}{V_{\kappa_n} \rho_n}. \tag{3.4}$$

Note that the data mechanism κ_n and the pricing strategy $s_n = \{\Pi_n, \pi_n\}$ will affect the MNO-n's threshold MU type $\sigma_n(\kappa_n, s_n)$. We define $\xi(\kappa)$ as in (3.5) to characterize the neutral MU type in subsequent analysis.

$$\xi(\kappa) \triangleq \frac{\rho_2 V_{\kappa_2}}{\rho_1 V_{\kappa_1}}. \tag{3.5}$$

3.2 User Subscription

Without loss of generality, we suppose that MNO-1 has an advantage in terms of $\rho_n V_{\kappa_n}$ among the two MNOs, i.e., $\xi(\kappa) < 1$. Hence we say MNO-1 is "stronger".

Definition 3.2 We define the neutral MU type $\tilde{\sigma}$ as the MU type who obtains the same payoff by subscribing to either MNO, i.e., $\bar{U}_1(\kappa_1, s_1, \tilde{\sigma}) = \bar{U}_2(\kappa_2, s_2, \tilde{\sigma})$.

$$\tilde{\sigma}(\sigma_1, \sigma_2) = \frac{\sigma_1 - \xi(\kappa) \cdot \sigma_2}{1 - \xi(\kappa)}, \quad (3.6)$$

where σ_1 and σ_2 are two MNOs' threshold MU types defined in Definition 3.1.

Under the data mechanism and pricing strategy, we analyze the threshold MU types and for MNOs. The market partition equilibrium can be categorized into three distinct scenarios.

1. The threshold MU type of MNO-1 is much larger than that of MNO-2, i.e., $(\sigma_1, \sigma_2) \in \Sigma_1$ where Σ_1 is

$$\Sigma_1 \triangleq \{(\sigma_1, \sigma_2) : \sigma_1 - \sigma_2 \geq (1 - \xi(\kappa))(\theta_{\max} - \sigma_2)\}. \quad (3.7)$$

The market share of MNO-2 corresponds to $\Phi_2^*(\kappa, s) = [\sigma_2, \theta_{\max}]$, while MNO-1 has a zero market share $\Phi_1^*(\kappa, s) = \varnothing$. Figure 3.1a illustrates the market partition.

2. The threshold MU type of MNO-1 is smaller than that of MNO-2, i.e., $(\sigma_1, \sigma_2) \in \Sigma_2$ where Σ_2 is

$$\Sigma_2 \triangleq \{(\sigma_1, \sigma_2) : \sigma_1 - \sigma_2 \leq 0\}. \quad (3.8)$$

The market share of MNO-1 corresponds to $\Phi_1^*(\kappa, s) = [\sigma_1, \theta_{\max}]$, while MNO-2 has a zero market share of $\Phi_2^*(\kappa, s) = \varnothing$. Figure 3.1b illustrates the market partition.

3. The threshold MU type of MNO-1 is slightly larger than that of MNO-2, i.e., $(\sigma_1, \sigma_2) \in \Sigma_3$ where Σ_3 is

$$\Sigma_3 \triangleq \{(\sigma_1, \sigma_2) : 0 < \sigma_1 - \sigma_2 < (1 - \xi(\kappa))(\theta_{\max} - \sigma_2)\}. \quad (3.9)$$

The market share of MNO-1 is $\Phi_1^*(\kappa, s) = [\tilde{\sigma}, \theta_{\max}]$, and the market share of MNO-2 is $\Phi_2^*(\kappa, s) = [\sigma_2, \tilde{\sigma}]$. Figure 3.1c illustrates the market partition.

Our study reveals two possible market equilibrium outcomes: coexistence or one-MNO-dominance, determined by the data mechanism and pricing strategy. It is important to note that the one-MNO-dominance outcome differs from a traditional monopoly. Even if an MNO holds zero market share, its presence may still influence the surviving MNO's strategic decisions. We explore this dynamic further in Sect. 3.3.1.

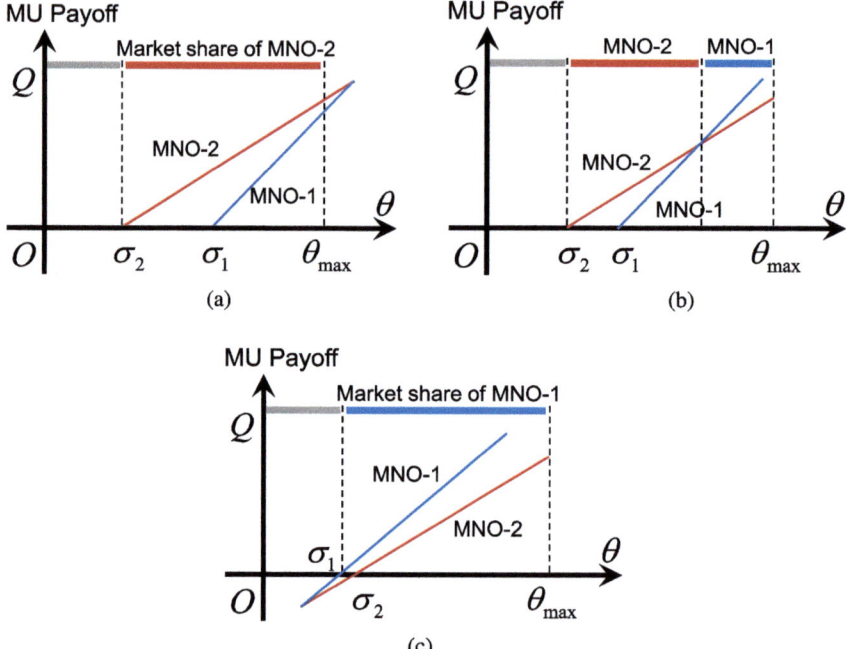

Fig. 3.1 The illustration of market partition in the duopoly market. (**a**) MNO-2 surviving (Σ_1). (**b**) MNO-1 surviving (Σ_2). (**c**) Share the market (Σ_3)

3.3 MNOs' Pricing Competition

In Stage II of MNOs' competition game, the MNOs would simultaneously determine their pricing strategies $s=\{s_1, s_2\}$, given $\kappa=\{\kappa_1, \kappa_2\}$ in Stage I and the market partition equilibrium in Stage III. Before analyzing the pricing competition, we first derive the two MNOs' profits (given the market partition equilibrium) as follows:

$$W_1(\kappa, \sigma) = \rho_1 V_{\kappa_1} \left[\sigma_1 - \frac{c_1}{\rho_1}\right]\left[1 - H\big(\tilde{\sigma}(\sigma_1, \sigma_2)\big)\right], \tag{3.10}$$

$$W_2(\kappa, \sigma) = \rho_2 V_{\kappa_2} \left[\sigma_2 - \frac{c_2}{\rho_2}\right]\left[H\big(\tilde{\sigma}(\sigma_1, \sigma_2)\big) - H(\sigma_2)\right], \tag{3.11}$$

where σ_1, σ_2, and $\tilde{\sigma}$ depend on the pricing strategies s and data mechanisms κ. Note that the MNOs' profits at the equilibrium of Stage III are uniquely determined by the data mechanism $\kappa = \{\kappa_1, \kappa_2\}$ and the threshold MU types $\sigma = \{\sigma_1, \sigma_2\}$. Moreover, (3.4) indicates that MNO-n can arbitrarily change the threshold MU type $\sigma_n(\kappa_n, s_n)$ by adjusting the pricing strategies $s_n = \{\Pi_n, \pi_n\}$. That is, the MNOs'

3.3 MNOs' Pricing Competition

price competition in Stage II can be equivalently formulated as the following threshold competition.

Game 3.1 *Given the data mechanism $\kappa = \{\kappa_1, \kappa_2\}$, the threshold competition of the two MNOs in Stage II could be formulated as the following game:*

- *Player: MNO-1 and MNO-2.*
- *Strategy: The MNO-n selects its threshold MU type $\sigma_n \in [c_n/\rho_n, \theta_{max}]$.*
- *Preference: The MNO-n gains the profit $W_n(\kappa, \sigma)$.*

In the following, we investigate the MNOs' best responses, and then find the equilibrium of Game 3.1. Before that, we introduce two notations as follows. Part of the best response analysis is related to the MNO-n's optimal threshold MU type $\sigma_n^{MP}(\kappa_n)$ given the data mechanism κ_n in the *monopoly* market. In the following, we will directly use $\sigma_n^{MP}(\kappa_n)$. Our subsequent analysis will demonstrate that the cost-to-QoS ratio c_n/ρ_n is a critical factor in determining MNOs' best responses. To streamline our notation, we introduce the following definition:

$$\psi_n \triangleq \frac{c_n}{\rho_n}, \ \forall n \in \{1, 2\}. \tag{3.12}$$

3.3.1 Best Response Analysis

Weaker MNO-2 We first investigate the best response of the weaker MNO-2. We first introduce the MNO-1's *winning* threshold $\theta_1^W \triangleq \psi_2$, *no-influence* threshold $\theta_1^N \triangleq \xi(\kappa)\sigma_2^{MP}(\kappa_2) + (1 - \xi(\kappa))\theta_{max}$, and *losing* threshold θ_1^L that satisfies

$$\frac{\theta_1^L - (1-\xi(\kappa))\theta_{max}}{\xi(\kappa)} - \frac{1 - H\left(\frac{\theta_1^L - (1-\xi(\kappa))\theta_{max}}{\xi(\kappa)}\right)}{h\left(\frac{\theta_1^L - (1-\xi(\kappa))\theta_{max}}{\xi(\kappa)}\right)} = \psi_2. \tag{3.13}$$

Let σ_1 denote MNO-1's threshold MU type, MNO-2 maximizes its profit $W_2(\kappa, \sigma)$ by choosing a threshold MU type $\sigma_2^*(\kappa, \sigma_1)$ as follows:

$$\sigma_2^*(\kappa, \sigma_1) = \begin{cases} \psi_2, & \text{if } \sigma_1 \in [0, \theta_1^W), \\ \hat{\sigma}_2, & \text{if } \sigma_1 \in [\theta_1^W, \theta_1^L), \\ \frac{\sigma_1 + (\xi(\kappa) - 1)\theta_{max}}{\xi(\kappa)}, & \text{if } \sigma_1 \in [\theta_1^L, \theta_1^N), \\ \sigma_2^{MP}(\kappa_2), & \text{if } \sigma_1 \in [\theta_1^N, \theta_{max}]. \end{cases} \tag{3.14}$$

Here $\hat{\sigma}_2$ solves $\hat{\sigma}_2 - \frac{H(\tilde{\sigma}(\sigma_1, \hat{\sigma}_2)) - H(\hat{\sigma}_2)}{\frac{\xi(\kappa)}{1-\xi(\kappa)}h(\tilde{\sigma}(\sigma_1, \hat{\sigma}_2)) + h(\hat{\sigma}_2)} = \psi_2$, where $h(\cdot)$ and $H(\cdot)$ represent the PDF and CDF of MUs' data valuation θ, respectively.

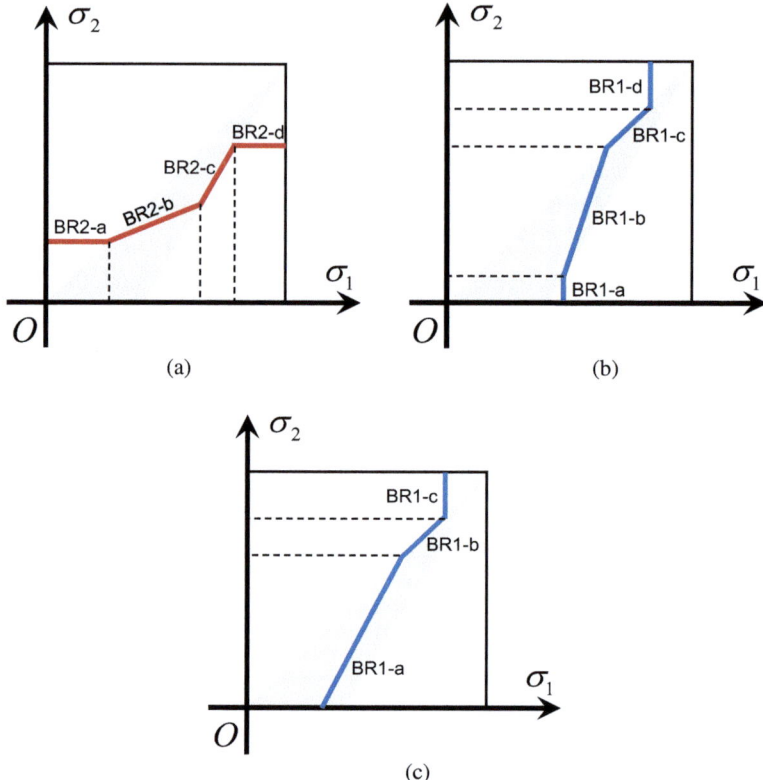

Fig. 3.2 Illustration of the best response for two MNOs. (**a**) Best response of MNO-2. (**b**) Best response of MNO-1. (**c**) Best response of MNO-1

These analytical results hold for any value distribution θ that satisfies the increasing failure rate (IFR) property. To provide concrete illustration, Fig. 3.2a demonstrates MNO-2's best response behavior under a uniform value distribution where $h(\theta) = 1/\theta_{\max}$. In the figure, the red line segments (labeled BR2-a through BR2-d) represent the optimal threshold strategy $\sigma_2^*(\kappa, \sigma_1)$. We now examine the operational significance of these four distinct response components:

- **BR2-a**: In this scenario, MNO-2 exits the competitive landscape, resulting in zero market acquisition. This represents a strategic withdrawal where the MNO ceases to actively compete for market share, i.e., $(\sigma_1, \sigma_2^*(\kappa, \sigma_1)) \in \Sigma_2$, if MNO-1's threshold MU type is smaller than its *winning* threshold, i.e., $\sigma_1 < \theta_1^W$.
- **BR2-b**: In this scenario, MNO-2 and MNO-1 reach a market-sharing equilibrium, partitioning the MU population base according to their respective competitive advantages, i.e., $(\sigma_1, \sigma_2^*(\kappa, \sigma_1)) \in \Sigma_3$, if MNO-1's threshold MU type is between its *winning* and *losing* thresholds, i.e., $\theta_1^W \leq \sigma_1 < \theta_1^L$.

3.3 MNOs' Pricing Competition

- **BR2-c**: In this scenario, MNO-2 achieves complete market dominance, reducing MNO-1's market share to zero in the equilibrium outcome, i.e., $(\sigma_1, \sigma_2^*(\kappa, \sigma_1)) \in \Sigma_1$, if MNO-1's threshold MU type is between its *losing* and *no-influence* thresholds, i.e., $\theta_1^L \leq \sigma_1 < \theta_1^N$.
- **BR2-d**: In this scenario, MNO-2 attains monopolistic market dominance, unilaterally determining its optimal threshold MU type independent of MNO-1's competitive presence, i.e., $\sigma_2^*(\kappa, \sigma_1) = \sigma_2^{MP}(\kappa_2)$, if MNO-1's threshold MU type is no smaller than its *no-influence* threshold, i.e., $\theta_1^N \leq \sigma_1 \leq \theta_{max}$.

Stronger MNO-1 We now analyze the best response dynamics of the stronger MNO-1 when reacting to the weaker MNO-2's market position. This asymmetric competition scenario reveals how power differentials influence strategic decision-making in duopoly markets. Similarly, we first define MNO-2's *winning* threshold $\theta_2^W \triangleq \frac{\psi_1 + (\xi(\kappa)-1)\theta_{max}}{\xi(\kappa)}$, *no-influence* threshold $\theta_2^N \triangleq \sigma_1^{MP}(\kappa_1)$, and *losing* threshold θ_2^L that satisfies

$$\theta_2^L - \frac{1-H(\theta_2^L)}{\frac{1}{1-\xi(\kappa)}h(\theta_2^L)} = \psi_1. \tag{3.15}$$

Under any value distribution θ satisfying the increasing failure rate (IFR) property, the lower threshold θ_2^L exists and is uniquely determined. Given MNO-2's threshold MU type σ_2, MNO-1 maximizes its profit $W_1(\kappa, \sigma)$ by the threshold MU type $\sigma_1^*(\kappa, \sigma_2)$, which satisfies

$$\sigma_1^*(\kappa, \sigma_2) = \begin{cases} \psi_1, & \text{if } \sigma_2 \in [0, \theta_2^W), \\ \hat{\sigma}_1, & \text{if } \sigma_2 \in [\theta_2^W, \theta_2^L), \\ \sigma_2, & \text{if } \sigma_2 \in [\theta_2^L, \theta_2^N), \\ \sigma_1^{MP}(\kappa_1), & \text{if } \sigma_2 \in [\theta_2^N, \theta_{max}], \end{cases} \tag{3.16}$$

where $\hat{\sigma}_1$ solves $\hat{\sigma}_1 - \frac{1-H(\tilde{\sigma}(\hat{\sigma}_1, \sigma_2))}{\frac{1}{1-\xi(\kappa)}h(\tilde{\sigma}(\hat{\sigma}_1, \sigma_2))} = \psi_1$.

While our analytical results hold for any value distribution θ satisfying IFR property, we employ a uniform distribution for concrete illustration of these findings. As shown in Fig. 3.2b, when MNO-1 has a relatively large cost-QoS ratio, i.e., $\psi_1 > (1-\xi(\kappa))\theta_{max}$, the corresponding insights are similar to Fig. 3.2a. In contrast, as shown in Fig. 3.2c, when MNO-1 has a small cost-QoS ratio, i.e., $\psi_1 \leq (1-\xi(\kappa))\theta_{max}$, MNO-2's *winning* threshold θ_2^W is always negative. This implies that MNO-1 maintains a strictly positive market share regardless of how small the MNO-2's threshold MU type σ_2 becomes. This outcome arises due to MNO-2's weaker competitive position. As shown in Fig. 3.2b, this configuration represents a degenerate case of the more general scenario depicted in Fig. 3.2c.

3.3.2 Equilibrium Analysis

Based on the above best responses, we study the threshold equilibrium of *Game 3.1*, denoted by $\boldsymbol{\sigma}^* = \{\sigma_1^*, \sigma_2^*\}$. Roughly speaking, Game 3.1 admits five distinct equilibrium types, characterized by the cost-QoS parameter pairs (ψ_1, ψ_2):

(1) In MNO-1's strong monopoly regime $(\psi_1, \psi_2) \in \Psi_1^{SM} = \{(\psi_1, \psi_2) : \psi_2 > \theta_2^N(\psi_1, \xi)\}$, MNO-2 exits the competitive market, thus MNO-1 can unilaterally decide its threshold MU type without considering the impact of MNO-2. Accordingly, MNO-2 obtains a zero market share, and the equilibrium is

$$\boldsymbol{\sigma}^*(\kappa) = \{\sigma_1^{MP}(\kappa_1), \psi_2\}. \tag{3.17}$$

(2) In MNO-1's weak monopoly regime $(\psi_1, \psi_2) \in \Psi_1^{WM} = \{(\psi_1, \psi_2) : \theta_2^L(\psi_1, \xi) < \psi_2 \leq \theta_2^N(\psi_1, \xi)\}$, MNO-2 still attempts to compete with MNO-1, but in vain. However, MNO-1 has to decide its threshold MU type considering MNO-2. Finally, MNO-2 obtains a zero market share and the equilibrium is

$$\boldsymbol{\sigma}^* = \{\psi_2, \psi_2\}. \tag{3.18}$$

(3) In coexistence regime $(\psi_1, \psi_2) \in \Psi^C = \{(\psi_1, \psi_2) : \psi_2 \leq \theta_2^L(\psi_1, \xi), \psi_1 \leq \theta_1^L(\psi_2, \xi)\}$. Both MNOs share the market, and the equilibrium $\boldsymbol{\sigma}^*(\kappa)$ solves

$$\begin{cases} \sigma_1^* - \dfrac{1 - H\left(\tilde{\sigma}(\sigma_1^*, \sigma_2^*)\right)}{\frac{1}{1-\xi(\kappa)} h\left(\tilde{\sigma}(\sigma_1^*, \sigma_2^*)\right)} = \psi_1, \\[2mm] \sigma_2^* - \dfrac{H\left(\tilde{\sigma}(\sigma_1^*, \sigma_2^*)\right) - H\left(\sigma_2^*\right)}{\frac{\xi(\kappa)}{1-\xi(\kappa)} h\left(\tilde{\sigma}(\sigma_1^*, \sigma_2^*)\right) + h\left(\sigma_2^*\right)} = \psi_2. \end{cases} \tag{3.19}$$

(4) In MNO-2's weak monopoly regime $(\psi_1, \psi_2) \in \Psi_2^{WM} = \{(\psi_1, \psi_2) : \theta_1^L(\psi_2, \xi) < \psi_1 \leq \theta_1^N(\psi_2, \xi)\}$, MNO-1 still attempts to compete with MNO-2, but in vain. MNO-2 has to decide its threshold MU type *considering* the impact of MNO-1. MNO-1 obtains a zero market share, and the equilibrium is

$$\boldsymbol{\sigma}^*(\kappa) = \left\{\psi_1, \frac{\psi_1 + (\xi(\kappa) - 1)\theta_{max}}{\xi(\kappa)}\right\}. \tag{3.20}$$

(5) In MNO-2's strong monopoly regime $(\psi_1, \psi_2) \in \Psi_2^{SM} = \{(\psi_1, \psi_2) : \psi_1 > \theta_1^N(\psi_2, \xi)\}$, MNO-1 gives up the competition, thus MNO-2 decides its threshold MU type without considering the impact of MNO-1. MNO-2 obtains a zero market, and the equilibrium is

$$\boldsymbol{\sigma}^*(\kappa) = \{\psi_1, \sigma_2^{MP}(\kappa_2)\}. \tag{3.21}$$

Having fully characterized the Stage II equilibrium threshold strategy $\sigma^*(\kappa)$ under data mechanism κ, we now proceed to analyze the MNOs' strategic mechanism selection in Stage I.

3.4 MNOs' Mechanism Competition

In Stage I, the two MNOs will decide their data mechanisms $\kappa = \{\kappa_1, \kappa_2\}$, considering the responses from Stage II and Stage III. We model the two MNOs' data mechanism selection as the following game.

Game 3.2 *The data mechanism selection game of the two MNOs in Stage I is as follows:*

- *Player: MNO-1 and MNO-2.*
- *Strategy: The MNO-n chooses its data mechanism κ_n from $\{T, R\}$.*
- *Preference: The MNO-n gains profit $W_n(\kappa)$.*

In Game 3.2, each MNO-n strategically selects its data mechanism $\kappa_n \in T, R$ to maximize profit $W_n(\kappa)$, accounting for the Stage II threshold equilibrium $\sigma^*(\kappa)$. To characterize the Nash equilibrium, we construct the complete payoff matrix in Table 3.1, enumerating all four possible mechanism combinations. Given the finite two-by-two structure, we employ exhaustive case analysis to verify equilibrium conditions for each strategy profile.

Throughout our analysis, we assume without loss of generality that MNO-1 provides superior average quality-of-service, i.e., $\rho_1 \geq \rho_2$. This natural ordering allows us to designate MNO-1 as the high-QoS provider and MNO-2 as the low-QoS provider in our duopoly framework. The cost structures (c_1, c_2) of both MNOs will critically influence the equilibrium outcomes. Our model explicitly accounts for the strategic interplay between these cost parameters and the QoS differentials, which jointly determine the market equilibrium.

3.4.1 Market Partition at Game 3.2 Equilibrium

We first introduce the three possible market partitions at a Game 3.2 equilibrium.

Table 3.1 MNOs' profits in Game 3.2

MNO-1: κ_1	MNO-2: κ_2 T	R
T	$W_1(T, T), W_2(T, T)$	$W_1(T, R), W_2(T, R)$
R	$W_1(R, T), W_2(R, T)$	$W_1(R, R), W_2(R, R)$

MNO-1 Surviving If MNO-1 experiences an extremely small cost, i.e., $c_1 < C_1^{Single}(\rho_1, \rho_2, c_2)$, then MNO-2 obtains zero market share.

MNO-2 Surviving If MNO-2 experiences an extremely small cost, i.e., $c_2 < C_2^{Single}(\rho_1, \rho_2, c_1)$, then MNO-1 obtains zero market share.

Coexistence If the costs of the two MNOs are comparable, i.e., $c_1 \geq C_1^{Single}(\rho_1, \rho_2, c_2)$ and $c_2 \geq C_2^{Single}(\rho_1, \rho_2, c_1)$, then they will share the market.

$$C_1^{Single}(\rho_1, \rho_2, c_2) \triangleq \rho_1 \left[\frac{c_2}{\rho_2} - \left(1 - \frac{\rho_2 V_T}{\rho_1 V_R}\right) \cdot \frac{1 - H\left(\frac{c_2}{\rho_2}\right)}{h\left(\frac{c_2}{\rho_2}\right)} \right], \quad (3.22)$$

$$C_2^{Single}(\rho_1, \rho_2, c_1) \triangleq \max \left\{ \rho_2 \left[\frac{c_1}{\rho_1} - \left(1 - \frac{\rho_1 V_T}{\rho_2 V_R}\right) \frac{1 - H\left(\frac{c_1}{\rho_1}\right)}{h\left(\frac{c_1}{\rho_1}\right)} \right], \right.$$
$$\left. c_1 - (\rho_1 - \rho_2)\theta_{max} - \rho_2 \cdot \frac{1 - H\left(\frac{c_1 - (\rho_1 - \rho_2)\theta_{max}}{\rho_2}\right)}{h\left(\frac{c_1 - (\rho_1 - \rho_2)\theta_{max}}{\rho_2}\right)} \right\}. \quad (3.23)$$

We now characterize the precise equilibrium mechanisms κ^* for each region, using the notation Na to denote non-applicable cases where:

- The equilibrium configuration $\kappa^* = (\kappa_1^*, \text{Na})$ indicates that MNO-1's optimal mechanism κ_1^* remains consistent regardless of MNO-2's choice, implying Game 3.2 admits two Nash equilibria (κ_1^*, T) and (κ_1^*, R).
- The equilibrium configuration $\kappa^* = (\text{Na}, \kappa_2^*)$ indicates that MNO-2's optimal mechanism choice κ_2^* is invariant to MNO-1's selection, implying Game 3.2 admits two Nash equilibria (T, κ_2^*) and (R, κ_2^*).

3.4.2 Single-MNO-Surviving

We now characterize the data mechanism equilibrium κ^* for market configurations where only one MNO achieves positive market share. This single-MNO-surviving equilibrium emerges when competitive dynamics lead to the complete exclusion of one MNO.

If $c_1 < C_1^{Single}(\rho_1, \rho_2, c_2)$, then the high-QoS MNO-1 obtains a positive market share, leaving a zero market share to the low-QoS MNO-2, which is independent of his choice. In this case, the data mechanism equilibrium is

$$\kappa^* = (\text{R}, \text{Na}). \quad (3.24)$$

3.4 MNOs' Mechanism Competition

Furthermore, we have

$$W_1(\text{R}, \text{T}) = W_1(\text{R}, \text{R}). \tag{3.25}$$

If $c_2 < C_2^{Single}(\rho_1, \rho_2, c_1)$, then the low-QoS MNO-2 obtains a positive market share, leaving a zero market share to the high-QoS MNO-1, which is independent of his choice. In this case, the data mechanism equilibrium is

$$\kappa^* = (\text{Na}, \text{R}). \tag{3.26}$$

Furthermore, we have

$$W_2(\text{R}, \text{R}) \leq W_2(\text{T}, \text{R}). \tag{3.27}$$

The above analysis reveals the impact of QoS advantage MNO-1 (i.e., $\rho_1 \geq \rho_2$): When the low-QoS MNO-2 obtains a zero market share, there is no effect on the high-QoS MNO-1 no matter its choice is T or R. That is, we have $W_1(\text{R}, \text{T}) = W_1(\text{R}, \text{R})$. When the high-QoS MNO-1 obtains a zero market share, it is possible to further reduce the profit of the low-QoS MNO-2 by choosing the rollover mechanism R. That is, we have $W_2(\text{R}, \text{R}) \leq W_2(\text{T}, \text{R})$.

3.4.3 Coexistence

We now examine the scenario where both MNOs achieve positive market shares at the equilibrium of Game 3.2, satisfying the conditions $c_1 \geq C_1^{Single}(\rho_1, \rho_2, c_2)$ and $c_2 \geq C_2^{Single}(\rho_1, \rho_2, c_1)$. To support our subsequent analysis, we first define the QoS-Flip phenomenon. Intuitively, the rollover mechanism can counteract MNO-2's initial QoS disadvantage. We formally define this behavior as follows:

Definition 3.3 The QoS-flip happens if the low-QoS MNO-2 attracts the high valuation MUs under the data mechanism $\kappa = \{\text{T}, \text{R}\}$.

In our subsequent equilibrium analysis for the coexistence scenario, we will characterize the conditions under which QoS-flip occurs. Under the coexistence equilibrium in Game 3.2, we identify two QoS thresholds, $\tilde{\rho} > \hat{\rho}$, which lead to three distinct possible outcomes for the equilibrium κ^*:

(1) When MNO-1 possesses a significant QoS advantage relative to MNO-2 (i.e., $0 < \rho_2 \leq \hat{\rho}$), then the equilibrium κ^* is (R, T) if MNO-2's cost is large, i.e., $c_2 > C_2^{Roll}(\rho_1, \rho_2, c_1)$; and (R, R) if MNO-2's cost is small, i.e., $c_2 \leq C_2^{Roll}(\rho_1, \rho_2, c_1)$.
(2) When MNO-1 possesses a small QoS advantage relative to MNO-2 (i.e., $\hat{\rho} < \rho_2 \leq \tilde{\rho}$), then the equilibrium κ^* is (R, T) if MNO-2's cost is large, i.e., $c_2 > C_2^{Roll}(\rho_1, \rho_2, c_1)$; and (R, R) if MNO-1's and MNO-2's costs are small

and comparable, i.e., $c_1 \leq C_1^{Roll}(\rho_1, \rho_2, c_2)$ and $c_2 \leq C_2^{Roll}(\rho_1, \rho_2, c_1)$; and (T, R) if MNO-1's cost is large, i.e., $c_1 > C_1^{Roll}(\rho_1, \rho_2, c_2)$.

(3) When MNO-1 possesses a negligible QoS advantage relative to MNO-2, i.e., $\tilde{\rho} < \rho_2 \leq \rho_1$, then the equilibrium κ^* is (R, T) if MNO-1 experiences a small cost, i.e., $c_1 < C_1^{Roll}(\rho_1, \rho_2, c_2)$; and {(R, T), (T, R)} if MNO-1's and MNO-2's costs are large and comparable, i.e., $c_1 \geq C_1^{Roll}(\rho_1, \rho_2, c_2)$ and $c_2 \geq C_2^{Roll}(\rho_1, \rho_2, c_1)$; and (T, R) if MNO-2's cost is small, i.e., $c_2 < C_2^{Roll}(\rho_1, \rho_2, c_1)$.

3.5 Summary

This chapter examines duopoly competition in telecommunications markets, focusing on MNOs' strategic adoption of rollover mechanisms and pricing decisions. We formulate the economic interactions between two competing MNOs and MUs as a three-stage game framework, accounting for heterogeneity in both operators' (quality-of-service and costs) and MUs' (data valuations). Using backward induction, we derive analytical characterizations of: (1) market segmentation in Stage III, (2) pricing equilibrium among MNOs in Stage II, and (3) equilibrium rollover mechanism strategies in Stage I.

Unlike monopoly markets—where mobile network operators (MNOs) can unconditionally boost profits by adopting rollover mechanisms—the strategic landscape in duopoly markets introduces nuanced competitive dynamics. Our game-theoretic analysis demonstrates that equilibrium mechanism selection depends critically on the interplay of cost structures and quality-of-service (QoS) differentiation. Specifically, we identify a competitive erosion effect: as operational costs rise or the rival MNO's efficiency improves, the high-QoS provider faces diminishing returns from rollover mechanisms and may phase them out to maintain profitability. Interestingly, this creates a strategic opening for the low-QoS MNO to leapfrog by adopting rollover services, thereby enhancing its market appeal. Consequently, the market equilibrium evolves along a non-monotonic trajectory-transitioning from an initial (Rollover, Traditional) configuration toward (Traditional, Rollover), though not necessarily reaching the latter state due to intermediate competitive rebalancing.

So far, we have understand the economic impact of the time flexibility in monopoly and competitive markets. Next, let us move on to see how the location flexibility affects the telecommunication market.

Chapter 4
Location Flexibility in Overseas Market

Abstract In this chapter, we move on to the day-pass data service with location flexibility. Specifically, we will analyze a MU's optimal data consumption and day-pass service configuration based on the location MU profile. We will also investigate how the MNO monetizes its day-pass service with the uncertainty of MU demand.

Keywords Day-pass data service · Location flexibility · Dynamic configuration

4.1 Market Model

We study the overseas markets where the MNO provides with the location-flexible day-pass data service. We consider a set $\mathcal{M} = \{1, 2, \ldots, M\}$ of months. Each month $m \in \mathcal{M}$ consists of a set $\mathcal{T}_m = \{1, 2, \ldots, T_m\}$ of days. Suppose that the overseas market consists of a set $\mathcal{N}_m = \{1, 2, \ldots, N_m\}$ of MUs in month m, who travel overseas in some days within $\mathcal{T}_m$. For the MNO, we investigate how to monetize the day-pass data service in the overseas market. For each MU in the overseas market, we investigate the joint flexibility configuration and data consumption problem.

4.1.1 Mobile Data Services

The domestic data service offered by the MNO can be represented using a three-part tariff structure $\{Q, \Pi, \pi\}$ [57]. The MU subscribes to a monthly plan by paying a fixed fee Π, which grants a data allowance of Q. Any usage beyond this cap incurs an additional charge of π per unit of data. However, this domestic plan does not apply when the MU travels abroad. MNO provides two options for overseas data services:

- *Pay-as-You-Go Roaming:* The traditional model charges MUs a per-unit fee $\hat{\pi}$ for data consumed overseas. Typically, this rate is higher than the domestic overage fee (i.e., $\hat{\pi} > \pi$).

- *Location-Flexibility:* This newer option allows MUs to pay a daily fee p to include their overseas data usage within their domestic data cap for that day.

It is important to note that long-term international data plans typically impose significant costs on MUs without frequent international travel needs. Furthermore, short-term international data plans exhibit economic characteristics similar to usage-based roaming options.

4.1.2 User Model

We model each MU $n \in \mathcal{N}_m$ from the perspective of the travel profile $l^n[m]$, the demand vector $\boldsymbol{a}^n[m]$, and the valuation vector $\boldsymbol{\theta}^n[m]$.

- We let $l_t^n \in \{0, 1\}$ denote whether MU $n \in \mathcal{N}_m$ stays in domestic locations (i.e., $l_t^n = 0$) or overseas locations (i.e., $l_t^n = 1$) in day $t \in \mathcal{T}_m$. Accordingly, represents the travel profile of MU n in the overseas market of month m.
- We let a_t^n denote the potential maximal data demand of MU $n \in \mathcal{N}_m$ in day $t \in \mathcal{T}_m$. The MU can meet the entire demand or part of the demand. Mathematically, we suppose that a_t^n is a random variable on the support $[0, \bar{a}]$.
- We let θ_t^n denote the valuation of MU $n \in \mathcal{N}_m$ if he satisfies his entire data demand (i.e., a_t^n) in day $t \in \mathcal{T}_m$. Mathematically, we suppose that θ_t^n is a random variable with the support of $[0, \bar{\theta}]$.

Based on the above discussion, each MU is modeled by the travel profile $l^n[m] = \left(l_t^n \in \{0, 1\} : \forall t \in \mathcal{T}_m\right)$, the demand vector $\boldsymbol{a}^n[m] = \left(a_t^n \in [0, \bar{a}] : \forall t \in \mathcal{T}_m\right)$, and the valuation vector $\boldsymbol{\theta}^n[m] = \left(\theta_t^n \in [0, \bar{\theta}] : \forall t \in \mathcal{T}_m\right)$.

User Decisions For each MU $n \in \mathcal{N}_m$, two key decisions should be made within each month m, i.e., flexibility configuration $\boldsymbol{x}^n[m]$ and data consumption $\boldsymbol{y}^n[m]$.

- *Flexibility Configuration:* The binary variable $x_t^n \in \{0, 1\}$ denotes the flexibility configuration of MU n in day t. If $x_t^n = 1$, then MU n configures location-flexibility in day t. The data consumption in day t will be deducted from the domestic data cap allocation Q^n. The monetary cost is p per month. The case of $x_t^n = 0$ means that MU n chooses to pay the roaming fee $\hat{\pi}$ in a usage-based manner. The flexibility configuration decisions of MU $n \in \mathcal{N}_m$ within month m is

$$\boldsymbol{x}^n[m] = \left(x_t^n \in \{0, 1\} : \forall t \in \mathcal{T}_m\right). \quad (4.1)$$

- *Data Consumption:* The variable $y_t^n \in [0, 1]$ denotes the fraction of data demand that MU n decides to satisfy in day t. Given the data demand a_t^n and the data consumption decision y_t^n, MU n will consume $a_t^n y_t^n$ amount of data, which generates utility $\theta_t^n y_t^n$ in day t. The data consumption decision of MU $n \in \mathcal{N}_m$ within month m is given by

$$\boldsymbol{y}^n[m] = \left(y_t^n \in [0, 1] : \forall t \in \mathcal{T}_m\right). \quad (4.2)$$

4.1 Market Model

MU Payoff The monthly payoff of an MU is defined as the difference between his gained utility and the corresponding monetary cost. The monetary cost usually includes the daily sunk cost and the monthly opportunistic cost. We elaborate the MU's utility, sunk cost, and opportunistic cost.

- *Utility* of the MU depends on the demand that is satisfied and the corresponding valuation. Given the demand a_t^n and the valuation θ_t^n, the consumption decision y_t^n leads to the utility $\theta_t^n y_t^n$.
- *Sunk cost* depends on the MU's choice between usage-based roaming (i.e., $x_t^n = 0$) and location-flexibility (i.e., $x_t^n = 1$) when the MU is at overseas locations (i.e., $l_t^n = 1$). The payment of roaming is $l_t^n(1 - x_t^n)\hat{\pi} a_t^n y_t^n$. The payment of location-flexibility is $l_t^n x_t^n p$, and the corresponding data consumption is $a_t^n y_t^n$. Nevertheless, the data consumption will be applied toward the monthly data cap allocation, potentially triggering overage charges if the cumulative usage exceeds the cap threshold. The MU's sunk cost in day t is

$$l_t^n \left[x_t^n p + (1 - x_t^n)\hat{\pi} a_t^n y_t^n \right]. \tag{4.3}$$

- *Opportunistic cost* depends on the overage fee for exceeding the monthly data cap Q. Recall that the data cap will cover the domestic data consumption $(1 - l_t^n)a_t^n y_t^n$ and the overseas configured consumption $l_t^n x_t^n a_t^n y_t^n$. Hence the *cap-consumption* of MU n in day t is

$$g_t^n(x_t^n, y_t^n) \triangleq (1 - l_t^n)a_t^n y_t^n + l_t^n x_t^n a_t^n y_t^n. \tag{4.4}$$

Accordingly, the opportunistic cost of MU $n \in \mathcal{N}_m$ in month m is given by

$$\pi \left[\sum_{t \in \mathcal{T}_m} g_t^n(x_t^n, y_t^n) - Q \right]^+, \tag{4.5}$$

where $[\cdot]^+ = \max(0, \cdot)$ and π is the per-unit overage fee.

Now we define the MU's *daily virtual payoff* in day t as follows

$$f_t^n\left(x_t^n, y_t^n\right) \triangleq \theta_t^n y_t^n - l_t^n \left[x_t^n p + (1 - x_t^n)\hat{\pi} a_t^n y_t^n \right]. \tag{4.6}$$

Accordingly, the *monthly payoff* of MU $n \in \mathcal{N}_m$ is given by

$$U_m^n\left(\mathbf{x}^n[m], \mathbf{y}^n[m]\right) \triangleq \sum_{t \in \mathcal{T}_m} f_t^n\left(x_t^n, y_t^n\right) - \pi \left[\sum_{t \in \mathcal{T}_m} g_t^n(x_t^n, y_t^n) - Q \right]^+. \tag{4.7}$$

Note that the subscription fee Π does not affect the decisions $(\mathbf{x}^n[m], \mathbf{y}^n[m])$ of MU $n \in \mathcal{N}_m$. Hence we omit it here.

MU Payoff Maximization Problem The payoff maximization of the MU is a joint flexibility configuration and data consumption (J-FCDC) problem. J-FCDC is an online optimization problem, since MU cannot know his demand profile and the valuation profile accurately.

4.1.3 MNO Revenue

The MNO benefits from the payment of the $|\mathcal{N}_m|$ MUs in the overseas market. The monthly payment of MU $n \in \mathcal{N}_m$ is given by

$$R_m^n(p) \triangleq p \sum_{t \in \mathcal{T}_m} l_t^n x_t^n + \hat{\pi} \sum_{t \in \mathcal{T}_m} l_t^n (1 - x_t^n) a_t^n y_t^n + \pi \left[\sum_{t \in \mathcal{T}_m} g_t^n(x_t^n, y_t^n) - Q \right]^+. \quad (4.8)$$

The revenue components in (4.8) can be interpreted as follows: (1) the first term represents the MU's payment for flexibility configuration, (2) the second term captures total roaming charges, and (3) the third term corresponds to overage fees incurred when exceeding data allowances. We exclude the fixed subscription revenue Π from our analysis since user decisions do not affect this income stream. Consequently, the MNO's aggregate revenue from overseas market segment $\mathcal{N}_m$ is given by:

$$R_m(p) \triangleq \sum_{n \in \mathcal{N}_m} R_m^n(p). \quad (4.9)$$

The MNO must establish its pricing strategy p under conditions of market uncertainty. This decision problem becomes particularly challenging due to the inherent information asymmetry in overseas markets.

4.2 Off-line Solution and Insights

To start with, we first investigate the off-line J-FCDC problem, and try to figure out the key insights. We then leverage the obtained insights to solve the online policy design. Now we consider a generic MU in a generic month and suppress the MU index n and the month index m.

4.2 Off-line Solution and Insights

4.2.1 Off-line Problem and Reformulation

If the MU can accurately predict his demand profile a and valuation profile θ, then the MU needs to solve the following off-line J-FCDC problem:

Problem 4.1 (Off-line J-FCDC)

$$\{x^*, y^*\} = \arg\max U(x, y) \tag{4.10a}$$

$$\text{s.t.} \ x_t \in \{0, 1\}, \ \forall t \in \mathcal{T}, \tag{4.10b}$$

$$y_t \in [0, 1], \ \forall t \in \mathcal{T}. \tag{4.10c}$$

Note that Problem 4.1 is a nonlinear mixed integer programming, which is NP-hard in general. There are two difficulties in the objective function $U(x, y)$: a piece-wise linear term and a product between x_t and y_t. We will address them by reformulating Problem 4.1 in the following two steps.

First, we linearize $x_t y_t$ by introducing a new variable w_t, which represents the fraction of the data demand a_t covered by the data cap *by configuring flexibility* in day t. Moreover, we introduce the following inequality constraints to ensure the equality constraint $w_t = x_t y_t$:

$$0 \leq w_t \leq x_t, \ \forall t \in \mathcal{T}, \tag{4.11a}$$

$$y_t + x_t - 1 \leq w_t \leq y_t, \ \forall t \in \mathcal{T}, \tag{4.11b}$$

where (4.11a) ensures $w_t = 0$ if $x_t = 0$, and (4.11b) ensures $w_t = y_t$ if $x_t = 1$. With the new variables $\boldsymbol{w} = (w_t \in [0, 1] : \forall t \in \mathcal{T})$, the MU's virtual payoff $f_t(\cdot)$ defined in (4.6) and the cap-consumption $g_t(\cdot)$ defined in (4.4) could be written as follows:

$$f_t(x_t, y_t, w_t) = \theta_t y_t - l_t \big[x_t p + (y_t - w_t)\hat{\pi} a_t\big], \tag{4.12a}$$

$$g_t(y_t, w_t) = (1 - l_t)a_t y_t + l_t a_t w_t. \tag{4.12b}$$

Second, to eliminate the piece-wise linear term, we introduce a new variable z, which denotes the data consumption exceeding the data cap Q. The new variable z satisfies the following conditions to ensure the equivalence:

$$z \geq \sum_{t \in \mathcal{T}} g_t(y_t, w_t) - Q, \tag{4.13a}$$

$$z \geq 0. \tag{4.13b}$$

Given the above reformulation, the MU's monthly payoff can be expressed as

$$U(\boldsymbol{x}, \boldsymbol{y}, \boldsymbol{w}, z) = \sum_{t \in \mathcal{T}} f_t(x_t, y_t, w_t) - \pi z, \qquad (4.14)$$

and we obtain Problem 4.2, which is equivalent to Problem 4.1.

Problem 4.2 (Reformulated Off-line J-FCDC)

$$\{\boldsymbol{x}^*, \boldsymbol{y}^*, \boldsymbol{w}^*, z^*\} = \arg\max U(\boldsymbol{x}, \boldsymbol{y}, \boldsymbol{w}, z) \qquad (4.15a)$$
$$s.t.\ (4.11),\ (4.13), \qquad (4.15b)$$
$$x_t \in \{0, 1\},\ \forall t \in \mathcal{T}, \qquad (4.15c)$$
$$y_t \in [0, 1],\ \forall t \in \mathcal{T}. \qquad (4.15d)$$

Problem 4.2 remains a mixed-integer linear program (MILP). By relaxing the integer constraints (4.15c)–(4.16c), we derive its linear programming counterpart, Problem 4.3. This relaxation serves primarily to reveal fundamental structural properties of our model, while our proposed online strategy operates without requiring such simplification.

Problem 4.3 (Relaxed Reformulated Off-line J-FCDC)

$$\{\boldsymbol{x}^\dagger, \boldsymbol{y}^\dagger, \boldsymbol{w}^\dagger, z^\dagger\} = \arg\max U(\boldsymbol{x}, \boldsymbol{y}, \boldsymbol{w}, z) \qquad (4.16a)$$
$$s.t.\ (4.11),\ (4.13), \qquad (4.16b)$$
$$x_t \in [0, 1],\ \forall t \in \mathcal{T}, \qquad (4.16c)$$
$$y_t \in [0, 1],\ \forall t \in \mathcal{T}. \qquad (4.16d)$$

We now derive the optimal solution to Problem 4.3, denoted by the quadruple $\{\boldsymbol{x}^\dagger, \boldsymbol{y}^\dagger, \boldsymbol{w}^\dagger, z^\dagger\}$. Through this derivation process, we will systematically uncover the fundamental principles and insights governing MUs' optimal decision-making behavior.

4.2.2 Key Insights of Solving Problem 4.3

Problem 4.3 falls into the category of linear programs. One can solve it based on the Karush-Kuhn-Tucker (KKT) conditions. What is important is to understand the MU's marginal value and marginal cost of consuming the data cap.

Marginal Cost For analytical clarity in subsequent discussion, we define λ and Λ as the Lagrangian multipliers corresponding to constraints (4.13a) and (4.13b), respectively. Drawing from economic optimization theory, λ represents the shadow

4.2 Off-line Solution and Insights

price—specifically, the marginal utility cost associated with consuming an additional unit of the data cap under the given constraints. For any λ satisfying the KKT conditions of Problem 4.3, we have

$$0 \le \lambda \le \pi. \tag{4.17}$$

We find that the non-negative shadow price λ is bounded above by the overage fee π. The equality $\lambda = \pi$ holds precisely when the monthly data cap Q is exhausted (i.e., when $Q \le \sum_{t \in \mathcal{T}} g_t(x_t, y_t, w_t)$). This relationship admits the following economic interpretation:

- If the data cap is large enough, i.e., $Q > \sum_{t \in \mathcal{T}} g_t(x_t, y_t, w_t)$, then no overage fee will be incurred. The marginal cost of using data cap is zero, i.e., $\lambda = 0$.
- If the data cap is small, i.e., $Q < \sum_{t \in \mathcal{T}} g_t(x_t, y_t, w_t)$, then the MU will incur overage fee. The marginal cost of using data cap is the same as the real overage fee, i.e., $\lambda = \pi$.
- If the data cap satisfies $Q = \sum_{t \in \mathcal{T}} g_t(x_t, y_t, w_t)$, then the marginal cost of consuming data cap falls into the range $(0, \pi)$.

The shadow price λ critically influences MUs' decision-making in two key dimensions: (1) flexibility configuration preferences x and (2) data consumption patterns y, w, and z. This relationship is bidirectional—while λ affects user decisions, these choices simultaneously determine the equilibrium value of λ. The system converges to a stationary equilibrium characterized by the optimal shadow price $\lambda^\dagger$ and corresponding decisions $x^\dagger$, $y^\dagger$, $w^\dagger$, and $z^\dagger$, which jointly solve Problem 4.2. Next, we analyze how λ governs MU behavior through its connection with the marginal value of data cap consumption.

Marginal Value The MU's marginal value (of consuming data cap) depends on the KKT conditions of Problem 4.3. Moreover, it is also related to the MU's characteristics. We first classify the MU's daily characteristics (l_t, a_t, θ_t) into three groups by defining three binary indicators as follows:

$$d_t \triangleq 1 - l_t, \quad \text{(Domestic state)}, \tag{4.18a}$$

$$\bar{o}_t \triangleq l_t \cdot \mathbb{I}(\theta_t \ge a_t \hat{\pi}), \quad \text{(Overseas high-value state)}, \tag{4.18b}$$

$$\underline{o}_t \triangleq l_t \cdot \mathbb{I}(\theta_t < a_t \hat{\pi}), \quad \text{(Overseas low-value state)}, \tag{4.18c}$$

where $\mathbb{I}(\cdot)$ denotes the indicator function. The physical meaning of (4.18) is as follows: $d_t = 1$ is a domestic state, $\bar{o}_t = 1$ is an overseas high-value state, and $\underline{o}_t = 1$ is an overseas low-value state. By definition, we have $d_t + \bar{o}_t + \underline{o}_t = 1$. Based on the above notations, the MU's *marginal value* (of consuming the data cap) in day t is defined as

$$V_t(p) \triangleq d_t \cdot \frac{\theta_t}{a_t} + \bar{o}_t \cdot \frac{\hat{\pi} a_t - p}{a_t} + \underline{o}_t \cdot \frac{\theta_t - p}{a_t}. \tag{4.19}$$

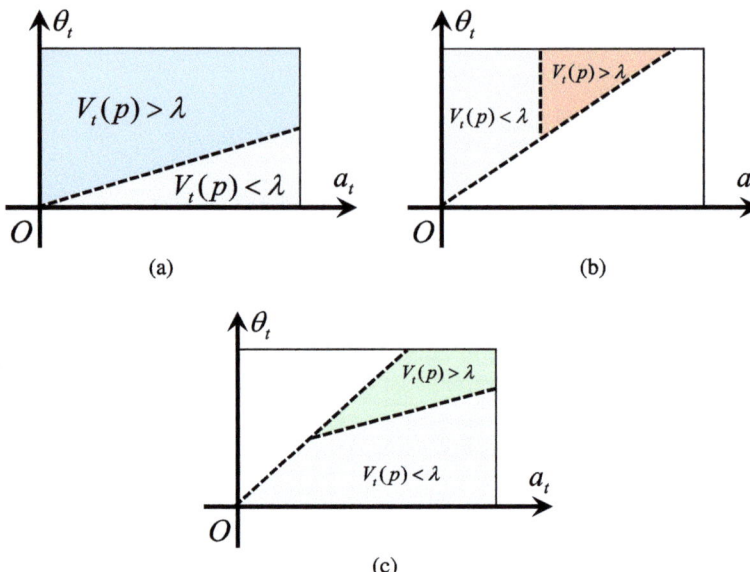

Fig. 4.1 An illustration of the offline insights. (**a**) Domestic state $d_t = 1$. (**b**) Overseas states $\bar{o}_t = 1$. (**c**) Overseas states $\underline{o}_t = 1$

The definition in (4.19) includes three cases. Based on (4.19), we are able to show that the marginal value satisfies $V_t(a_t, \theta_t, p) \leq \bar{v}$ and the upper bound $\bar{v}$ is

$$\bar{v} \triangleq \max\left(v, \hat{\pi}\right). \tag{4.20}$$

Now we elaborate the intuition of the marginal value.

- In domestic state $d_t = 1$, MUs exhibit two characteristic behaviors: (1) they never configure flexibility (i.e., $x_t = 0$), and (2) all data consumption is deducted from their monthly cap. The decision to fully satisfy demand (i.e., $y_t = 1$) occurs when the net marginal utility condition $V_t(p) > \lambda$ holds, which is mathematically equivalent to $\theta_t - \lambda a_t > 0$. This inequality indicates that the marginal benefit exceeds the opportunity cost of data cap consumption. As visualized in Fig. 4.1a, the blue region represents precisely these parameter combinations where the net value condition is satisfied for domestic states.
- In overseas high-value state $\bar{o}_t = 1$, MUs will always satisfy the entire data demand, since the valuation θ_t is larger than the roaming cost according to (4.18b). Consequently, the MU tends to pay for the usage-based roaming fee or access the monthly data cap by configuring location flexibility. Note that the inequality $V_t(p) > \lambda$ indicates $\hat{\pi} a_t > \lambda a_t + p$, which means that location flexibility incurs less monetary cost than the usage-based roaming fee. In Fig. 4.1b, the overseas high-value state $\bar{o}_t = 1$ is denoted by the region above the dash line and the red region means $V_t(p) > \lambda$.

4.2 Off-line Solution and Insights

- In the overseas low-value state $\underline{o}_t = 1$, the MU will avoid roaming fee according to (4.18c). Specifically, the MU either fully restricts the data demand or tends to access the data cap by configuring location flexibility. The inequality $V_t(p) > \lambda$ is equivalent to $\theta_t - p - \lambda a_t > 0$, which shows that location flexibility yields positive payoff. In Fig. 4.1c, the overseas low-value state $\underline{o}_t = 1$ is denoted by the region below the dash line and the green region means $V_t(p) > \lambda$.

When $V_t(p) > \lambda$, the marginal benefit exceeds the marginal cost of consuming the data allowance. Consequently, under these circumstances, the MU has an incentive to utilize the data cap. Building on this analysis, we formally define cap-consumption at shadow price λ as follows:

$$A(p, \lambda, l, a, \theta) \triangleq \sum_{t=1}^{T} a_t \cdot \mathbb{I}\left(V_t(p) > \lambda\right). \quad (4.21)$$

Note that $A(\lambda, p, l, a, \theta)$ is step-wise and weakly decreasing in λ. For notation simplicity, we let $\lambda^\star$ denote the minimal marginal cost that leads to zero overage fee. It is formally defined as

$$\lambda^\star \triangleq \arg\min_{t \in \mathcal{T}} V_t(p)$$
$$\text{s.t. } A(p, V_t(p), l, a, \theta) \leq Q, \quad (4.22)$$

As we will see later, we will leverage $\lambda^\star$ to solve Problem 4.3 analytically.

4.2.3 Solution of Problem 4.3

We present the solution $(x_t^\dagger, y_t^\dagger, w_t^\dagger)$ and $z^\dagger$ of Problem 4.3 in Theorem 4.1. The major rationale is to identify $\lambda^\dagger$, i.e., the optimal value of the shadow price.

Theorem 4.1 *The shadow price* $\lambda^\dagger = \mathcal{P}_{[0,\pi]}(\lambda^\star)$ *together with* $(x^\dagger, y^\dagger, w^\dagger)$ *in (4.23) and* $z^\dagger$ *in (4.24) satisfies the KKT conditions of Problem 4.3.*

$$\left(x_t^\dagger, y_t^\dagger, w_t^\dagger\right) = \begin{cases} (0,0,0)\left(d_t + \underline{o}_t\right) + (0,1,0)\overline{o}_t, & \text{if } V_t(p) < \lambda^\dagger, \\ (0,1,0)d_t + (1,1,1)\left(\underline{o}_t + \overline{o}_t\right), & \text{if } V_t(p) > \lambda^\dagger, \\ (0,\delta,0)d_t + (\delta,\delta,\delta)\underline{o}_t + (\delta,1,\delta)\overline{o}_t, & \text{if } V_t(p) = \lambda^\dagger, \end{cases}$$
$$(4.23)$$

$$z^\dagger = \left[A(p, \lambda^\dagger, l, a, \theta) - Q\right]^+. \quad (4.24)$$

Moreover, δ in (4.23) is given by

$$\delta = \mathcal{P}_{[0,1]}\left(\frac{Q - A(p, \lambda^\dagger, \boldsymbol{l}, \boldsymbol{a}, \boldsymbol{\theta})}{\sum_{t=1}^{T} a_t \cdot \mathbb{I}\left(V_t(p) = \lambda^\dagger\right)}\right), \tag{4.25}$$

where $\mathcal{P}_{[0,1]}(\cdot)$ represents the projection to the range $[0, 1]$.

Theorem 4.1 has three possible structures, which depend on the MU's characteristics and the data cap). We summarize the three cases as follows

$$\begin{aligned}
\Omega_H &\triangleq \{(\boldsymbol{l}, \boldsymbol{a}, \boldsymbol{\theta}) : A(p, \pi, \boldsymbol{l}, \boldsymbol{a}, \boldsymbol{\theta}) > Q\}, \\
\Omega_L &\triangleq \{(\boldsymbol{l}, \boldsymbol{a}, \boldsymbol{\theta}) : A(p, 0, \boldsymbol{l}, \boldsymbol{a}, \boldsymbol{\theta}) < Q\}, \\
\Omega_M &\triangleq \{(\boldsymbol{l}, \boldsymbol{a}, \boldsymbol{\theta}) : A(p, \pi, \boldsymbol{l}, \boldsymbol{a}, \boldsymbol{\theta}) \leq Q \leq A(p, 0, \boldsymbol{l}, \boldsymbol{a}, \boldsymbol{\theta})\},
\end{aligned} \tag{4.26}$$

and introduce the differences among the three cases.

1. The case of $(\boldsymbol{l}, \boldsymbol{a}, \boldsymbol{\theta}) \in \Omega_H$ represents that the MU has a heavy data demand. Accordingly, we have $A(p, \lambda^\dagger, \boldsymbol{l}, \boldsymbol{a}, \boldsymbol{\theta}) > Q$, since $\lambda^\star > \pi$ and $\lambda^\dagger = \pi$. In this case, $z^\dagger > 0$ and $\delta = 0$. That is, the heavy demand requires the MU to run out the data cap and incur overage fee.
2. The case of $(\boldsymbol{l}, \boldsymbol{a}, \boldsymbol{\theta}) \in \Omega_L$ represents that the MU has a light data demand. Accordingly, we have $A(p, \lambda^\dagger, \boldsymbol{l}, \boldsymbol{a}, \boldsymbol{\theta}) < Q$, since $\lambda^\star < 0$ and $\lambda^\dagger = 0$. In this case, $z^\dagger = 0$ and $\delta = 1$. That is, the MU has some data cap left.
3. The case of $(\boldsymbol{l}, \boldsymbol{a}, \boldsymbol{\theta}) \in \Omega_M$ represents that the MU has a medium data demand. Accordingly, we have $A(p, \lambda^\dagger, \boldsymbol{l}, \boldsymbol{a}, \boldsymbol{\theta}) \leq Q$, since $\lambda^\star = \lambda^\dagger \in (0, \pi)$. In this case, $z^\dagger = 0$ and $\delta \in (0, 1)$. The fraction value $\delta \in (0, 1)$ ensures that the MU would run out the data cap and stop consuming data.

4.3 MU's Online Strategy

We focus on the MU's online J-FCDC problem and proposes a strategy with provably performance. Recall that each MU should decide (x_t, y_t) in each day t with the goal of improving the monthly payoff defined as follow

$$U(\boldsymbol{x}, \boldsymbol{y}) = \sum_{t=1}^{T} f_t(x_t, y_t) - \pi \left[\sum_{t=1}^{T} g_t(x_t, y_t) - Q\right]^+, \tag{4.27}$$

4.3 MU's Online Strategy

where $f_t(x_t, y_t)$ is the MU's virtual payoff defined in (4.6) and $g_t(x_t, y_t)$ is the cap-consumption defined in (4.4). That is, MU's online J-FCDC problem is

$$\max_{x,y} \quad U(x, y) \tag{4.28}$$
$$s.t. \quad x_t \in \{0, 1\}, \ y_t \in [0, 1], \ \forall t \in \{1, 2, \ldots, T\}.$$

Although the online J-FCDC problem in (4.28) shares some similarities with the classic Lyapunov optimization framework and the online optimization framework, there are significant differences.

- The Lyapunov optimization framework aims to maximize the time-average reward while stabilizing the (virtual) queues (Theorem 4.2 in [58]). A control parameter tunes the trade-off between the reward maximization and the queue length. The J-FCDC problem in (4.28) has a virtual queue with the dynamics $Q(t+1) = Q(t) + g_t(x_t, y_t) - Q/T$. However, the overage payment in the objective function of (4.28) is piece-wise linear and is not decomposable over time.
- The online optimization framework tackles with the sequential decision-making problems, aiming to minimize regret. Moreover, it is usually assumed that the structure of the objective is additive [59]. In this setting, the online gradient decent (OGD) achieves sub-linear regret under the convexity property. However, J-FCDC problem in (4.28) has a piece-wise linear term related to all decisions.

As the analysis shows, neither framework above is suitable for our problem In the following, we leverage the insights discussed in Sect. 4.2 to solve this problem. First of all, we define the following Lagrangian function:

$$L_t(x_t, y_t, \lambda) = f_t(x_t, y_t) - \lambda \cdot \left[g_t(x_t, y_t) - \frac{Q}{T} \right], \ \forall t \in \mathcal{T}, \tag{4.29}$$

where $\lambda \in [0, \pi]$ corresponds to the marginal cost or shadow price mentioned in Sect. 4.2. Particularly, we note that

$$\sum_{t=1}^{T} L_t(x_t, y_t, \lambda) = \sum_{t=1}^{T} f_t(x_t, y_t) - \lambda \left(\sum_{t=1}^{T} g_t(x_t, y_t) - Q \right). \tag{4.30}$$

Comparing (4.30) and the objective of (4.28) indicates that if one minimizes $\sum_{t=1}^{T} L_t(x_t, y_t, \lambda)$ over $\lambda \in [0, \pi]$, then we obtain the MU's monthly payoff $U(x, y)$. That is, we have

$$\min_{\lambda \in [0,\pi]} \sum_{t=1}^{T} L_t(x_t, y_t, \lambda) = \min_{\lambda \in [0,\pi]} \sum_{t=1}^{T} f_t(x_t, y_t) - \lambda \cdot \left(g_t(x_t, y_t) - \frac{Q}{T} \right)$$

$$= \sum_{t=1}^{T} f_t(x_t, y_t) + \min_{\lambda \in [0,\pi]} \lambda \sum_{t=1}^{T} \left(\frac{Q}{T} - g_t(x_t, y_t) \right)$$

$$= \sum_{t=1}^{T} f_t(x_t, y_t) + \min_{\lambda \in [0,\pi]} \lambda \left(Q - \sum_{t=1}^{T} g_t(x_t, y_t) \right)$$

$$= \sum_{t=1}^{T} f_t(x_t, y_t) - \pi \left[\sum_{t=1}^{T} g_t(x_t, y_t) - Q \right]^+,$$

$$= U(x, y), \quad \forall\, (x, y). \tag{4.31}$$

The equality (4.31) indicates that the augmented Lagrangian functions are closely connected to the MU's monthly payoff. It enables us to decompose the piecewise linear term (i.e., the term related to the overage fee) in the online strategy design. Algorithm 2 introduces the online strategy $\mathfrak{S}$. The strategy $\mathfrak{S}$ will generate a sequence shadow price $\tilde{\lambda}$, which guides the MU to determines the flexibility configuration $\tilde{x}$ and data consumption $\tilde{y}$. It mainly comprises the following two steps in each day t.

- **Line 3:** We obtain the optimal decisions $\tilde{x}t$ and $\tilde{y}t$ by maximizing the augmented Lagrangian $L_t(x, y, \tilde{\lambda}t)$ given the current shadow price estimate $\tilde{\lambda}t$. This approach emulates the optimality conditions from our offline KKT analysis, but with a crucial distinction: in the online setting, the shadow price $\tilde{\lambda}_t$ varies dynamically across time periods. This temporal variation in opportunity costs leads to inevitable performance degradation compared to the offline benchmark solution.
- **Line 4:** We update the shadow price for the next day (i.e., $\tilde{\lambda}_{t+1}$) based on the relationship between the actual cap-consumption amount $g_t(\tilde{x}_t, \tilde{y}_t)$ and the daily average data cap Q/T. Moreover, this process involves an appropriate projection to $[0, \min(\pi, \bar{v})]$. Note that this step essentially imitates the derivation of the optimal shadow price in the off-line KKT analysis. Due to the causal nature of online decision-making where future data consumption patterns are unavailable, we update the shadow price by comparing only the current cap utilization $g_t(\tilde{x}_t, \tilde{y}_t)$ against the pro-rated daily allocation Q/T. This myopic adjustment, while suboptimal relative to the offline oracle that knows future consumption, provides a tractable online approximation of the ideal KKT conditions.

4.3 MU's Online Strategy

Algorithm 2: User's Online Strategy $\mathfrak{S}$

Input : Initial l, a, and θ.
Output: $(\tilde{x}, \tilde{y})$ and $\tilde{\lambda}$.
1 **Initial** $\tilde{\lambda}_1 = 0$ and step size $\eta = \{\eta_t, \forall t \in \mathcal{T}\}$.
2 **for** $t = 1$ to T **do**
3   *Decide* $(\tilde{x}_t, \tilde{y}_t)$ based on $\tilde{\lambda}_t$ as follows:

$$(\tilde{x}_t, \tilde{y}_t) = \arg\max_{x,y} L_t(x, y, \tilde{\lambda}_t) \qquad (4.32)$$
$$\text{s.t. } x \in \{0, 1\},\ y \in [0, 1].$$

4   *Update* $\tilde{\lambda}_{t+1}$ according to

$$\tilde{\lambda}_{t+1} = \mathcal{P}_{[0,\min(\pi,\bar{v})]}\left(\tilde{\lambda}_t - \eta_t \left[\frac{Q}{T} - g_t(\tilde{x}_t, \tilde{y}_t)\right]\right). \qquad (4.33)$$

While Algorithm 2 bears some resemblance to gradient-based methods, its operation is fundamentally online in nature. To assess the efficacy of strategy $\mathfrak{S}$, we employ a performance metric based on the monthly payoff gap, defined as:

$$G(\mathfrak{S}) \triangleq U(x^*, y^*) - U(\tilde{x}, \tilde{y}), \qquad (4.34)$$

where (x^*, y^*) is the offline optimal solution of Problem 4.1. As we will see later, the *cap-consumption fluctuation* and the *demand-supply divergence* will affect $G(\mathfrak{S})$.

- Given the MU's offline optimal decision (x^*, y^*), the leftover quota in day t (with respect to the daily average cap) $r_t^* = \frac{Q}{T} - g_t(x_t^*, y_t^*)$. We let $\bar{r}^* = \sum_{t=1}^{T} r_t^*/T$, and define the *cap-consumption fluctuation* in the first t days as follows:

$$\phi_t \triangleq \left|\bar{r}^* t - \sum_{k=1}^{t} r_k^*\right|, \qquad (4.35)$$

which quantifies the absolute difference between the cumulative remaining quota and the average leftover quota over the first t days. The maximal cap-consumption fluctuation within a month is defined as

$$\Phi \triangleq \max_{1 \leq t \leq T} \phi_t. \qquad (4.36)$$

In the special case of constant daily data consumption (where $g_t(x_t^*, y_t^*)$ remains invariant across all time periods t), the cap-consumption fluctuation becomes identically zero (i.e., $\phi_t = 0$, $\forall t$).

- For each time period t, given the random demand realization $a_t \in [0, \bar{a}]$, the *daily demand-supply mismatch* quantifies the deviation from the prescribed daily average capacity $\xi_t \triangleq \frac{Q}{T} - a_t$. Thus the *maximal demand-supply divergence* is

$$\Xi \triangleq \max\left(\left|\frac{Q}{T} - \bar{a}\right|, \frac{Q}{T}\right). \tag{4.37}$$

We summarize the performance of our proposed strategy $\mathfrak{S}$ in Theorem 4.2.

Theorem 4.2 *Suppose that the step size is $\eta_t = \frac{\min(\pi, \bar{v})}{\Xi\sqrt{T}}$, $\forall t \in \mathcal{T}$, the output $(\tilde{x}, \tilde{y})$ of Algorithm 2 satisfies*

$$G(\mathfrak{S}) \leq \min(\pi, \bar{v}) \cdot (\Xi + \Phi)\sqrt{T}, \tag{4.38}$$

where $\bar{v}$ defined in (4.20) is the MU's maximal marginal value, Ξ is the maximal demand-supply divergence, and Φ is the maximal cap-consumption fluctuation.

Theorem 4.2 characterizes three key determinants of the algorithm's performance:

Cost-Value Tradeoff Proposition 1 establishes that the marginal cost of data cap consumption lies in $[0, \pi]$. When the overage fee π dominates the user's marginal value $V_t(p)$, consumption decreases. Consequently, the performance gap $G(\mathfrak{S})$ scales linearly with $\min(\pi, \bar{v})$, where $\bar{v}$ bounds the marginal value.

Adaptive Marginal Cost Dynamics The strategy $\mathfrak{S}$ dynamically adjusts the instantaneous marginal cost using historical observations. The discrepancy between this adapted cost and the ex-post optimal cost depends on demand variability (i.e., maximal demand-supply divergence Ξ) and consumption patterns (i.e., maximal cap-consumption fluctuation Φ). Thus, $G(\mathfrak{S})$ grows linearly with both Ξ and Φ.

Temporal Convergence:] Over the T-day horizon, the gradient-based update rule (Eq. (4.33)) ensures the regret follows a square-root scaling law: $\mathcal{O}(\sqrt{T})$.

4.4 Summary

This chapter presents a comprehensive investigation of location-based flexibility enabled by day-pass data services in overseas markets, addressing the critical challenges posed by demand uncertainty and geographical mobility. Our study establishes a novel analytical framework to examine how MUs navigate the complex decision-making process involving: (1) dynamic flexibility configuration selection, and (2) adaptive data consumption strategies while roaming abroad. The inherent stochasticity in both data demand patterns and location transitions creates a unique online optimization problem that requires innovative solution approaches.

4.4 Summary

Our methodological approach proceeds through two key phases of analysis. First, we establish a foundational understanding by examining the offline version of this problem, where complete information about future demand and mobility patterns is assumed to be known a priori. We find that the optimal policy exhibits a threshold structure based on the MU's marginal valuation of data consumption. Moreover, the marginal cost of depleting the monthly data cap creates non-linear tradeoffs in service selection. Location-dependent value functions reveal how geographical factors influence optimal decision making. Second, building on these offline insights, we then develop a practical online policy designed for real-world implementation, where MUs must make decisions without future knowledge. Our proposed solution incorporates adaptive threshold mechanisms that respond to observed usage patterns, and balances immediate connectivity needs against future requirements through dynamic programming.

Chapter 5
User-identity Flexibility in Data-trading Market

Abstract While location flexibility addresses spatial constraints, user-identity flexibility through data trading creates new economic opportunities by enabling peer-to-peer resource sharing. In this chapter, we focus on the data-trading service, and analyze mobile MUs' optimal trading strategy and the MNO's revenue-maximizing trading prices. Moreover, we will also investigate how the rollover data service affects the equilibrium between mobile MUs and the MNO. This chapter will unveil the economic viability of integrating time flexibility and user-identity flexibility.

Keywords Data trading service · User-identity flexibility · Trading market

5.1 Market Model

We analyze a telecommunications market comprising a set of mobile users (MUs), denoted as $\mathcal{N} = \{1, 2, .., N\}$, subscribed to a mobile network operator (MNO). The MNO provides a three-part tariff data plan, enhanced with rollover and data trading features. The system operates in discrete time slots indexed by $t \in \{1, 2, 3, \ldots\}$, where each slot involves two key processes: (1) the MNO determines the trading prices, and (2) each MU $n \in \mathcal{N}$ takes a trading action (how much to buy or sell). At the end of each billing cycle (e.g., a month), data rollover is applied based on the MUs' cumulative data consumption and trading activities during that period.

5.1.1 Wireless Data Services

Mobile Data Plan A mobile data plan is defined by a tuple $\mathcal{T} = \{Q, \Pi, \pi\}$. The MU pays a *monthly subscription fee* Π for the data consumption up to the *data cap* Q. The MU pays the *overage fee* π for unit data consumption exceeding the data cap. In practice, MUs typically commit to a long-term contract (e.g., 1 or 2 years) with the MNO for a chosen plan. We analyze a period spanning M months, with each month divided into K time slots (e.g., $K = 30$ for daily slots). We denote $m \in \{1, 2, \ldots, M\}$ as the m-th month and $k \in \{1, 2, \ldots, K\}$ as the k-th time slot in a particular month. Given the time slot (m, k), the absolute time is $t = K(m-1)+k$. For simplicity, we interchangeably use (m, k) and t to reference time slots.

Rollover Mechanism The rollover feature enables MUs to transfer unused data from the previous month into the current month. However, implementations of rollover vary based on how consumed data is prioritized between rollover data (leftover from the previous month) and current monthly data allowance. In our model, we adopt the following rules:

- Rollover data is used before the current month's allocated data cap.
- Any remaining rollover data expires at the end of the current month and cannot be carried forward further.

Data Trading Market Within the data trading market, MUs can participate as either sellers of surplus data or buyers seeking additional data. MNO dynamically sets the selling price p_t^s and the buying price p_t^b in each time slot t. These prices are determined based on the aggregate supply from MU sellers and the aggregate demand from MU buyers. This chapter focuses on the MNO's optimal pricing strategy to maximize revenue. Formally, in each slot t, the MNO selects the price vector $\boldsymbol{p}t = p_t^s, pt^b$ to optimize its revenue. A key result of our analysis reveals that the revenue-maximizing prices naturally achieve market equilibrium that ensures complete clearance of the data trading market in each time period.

Implementation of Data Rollover and Trading The introduction of both rollover and data trading services presents new operational considerations for MNOs. Under this framework, MUs manage two distinct data allowances: a short-term cap (valid only for the current month and expiring at month-end) and a long-term cap (the base Q allocation that can roll over to subsequent months). Data trading directly impacts these allowances through specific policy constraints: purchased data receives top consumption priority but cannot roll over (becoming short-term), while when selling data, MUs must first liquidate short-term allowances (comprising both rollover data from the previous month and newly purchased data) before accessing long-term allocations. This tiered system—where short-term data is always consumed and sold before long-term data—optimizes flexibility while maintaining clear expiration protocols.

5.1.2 MUs' Decisions

Each MU n's data plan is characterized by the tuple $\{Q^n, \Pi^n, \pi\}$, where Q^n represents the data cap, Π^n denotes the subscription fee, and π is the uniform overage fee common across all plans offered by the MNO [10]. During time slot t, we track two state variables for each MU: e_t^n (short-term data volume) and q_t^n

5.1 Market Model

(long-term data volume), with the constraint $q_t^n \leq Q^n$ since Q^n serves as the upper bound for long-term data. These variables are aggregated across all MUs $\mathcal{N}$ into system-wide vectors $\boldsymbol{e}_t = \{e_t^n : n \in \mathcal{N}\}$ and $\boldsymbol{q}_t = \{q_t^n : n \in \mathcal{N}\}$. The temporal dynamics within each slot t are as follows. First of all, the MNO determines the prices, denoted by $\boldsymbol{p}_t = \{p_t^s, p_t^b\}$. Second, each MU $n \in \mathcal{N}$ checks the leftover data volume (e_t^n, q_t^n) as well as the trading prices $\boldsymbol{p}_t = \{p_t^s, p_t^b\}$. Third, each MU n determines the trading action, and then the leftover data becomes $(\bar{e}_t^n, \bar{q}_t^n)$. Forth, given the data consumption, the leftover data volume decreases to $(\hat{e}_t^n, \hat{q}_t^n)$ at the end of slot t. In the following, we introduce MUs' trading actions and data consumptions. We define $(\cdot)^+ = \max\{\cdot, 0\}$ and $(\cdot)^- = \min\{\cdot, 0\}$ for brevity.

Trading Action Let a_t^n denote the trading action of MU n in time slot t. The trading action has three cases.

- Buying data corresponds to $a_t^n > 0$. That is, MU n pay $p_t^b a_t^n$ to buy a_t^n amount of data. The short-term data volume becomes $\bar{e}_t^n = e_t^n + a_t^n$, while the long-term data volume remains the same, i.e., $\bar{q}_t^n = q_t^n$.
- Selling data corresponds to $a_t^n < 0$. MU n gains $p_t^s |a_t^n|$ by selling $|a_t^n|$ amount of data. The short-term data is sold first, thus becomes $\bar{e}_t^n = (e_t^n + a_t^n)^+$. If the short-term data is not enough to, then the MU would further sell the long-term data, which becomes $\bar{q}_t^n = q_t^n + (e_t^n + a_t^n)^-$.
- No trading action corresponds to $a_t^n = 0$. In this case, the data volume of the MU remains the same, and we have $\bar{e}_t^n = e_t$ and $\bar{q}_t^n = q_t^n$.

Consequently, the remaining short-term and long-term data volumes for MU n following data trading can be expressed as:

$$\begin{cases} \bar{e}_t^n = (e_t^n + a_t^n)^+, \\ \bar{q}_t^n = q_t^n + (e_t^n + a_t^n)^-. \end{cases} \quad (5.1)$$

We let $\bar{z}_t^n$ represent the total data volume after MU n conducts the trading action a_t^n. Mathematically, $\bar{z}_t^n$ is given by

$$\bar{z}_t^n = \bar{e}_t^n + \bar{q}_t^n = e_t^n + q_t^n + a_t^n. \quad (5.2)$$

Data Consumption For each MU n, we model the data consumption x_t^n in time slot t as a random variable with probability density function $f_n(\cdot)$. This stochastic approach captures the inherent uncertainty in user consumption patterns, as MUs cannot precisely predict their future data usage. The distribution can be estimated from historical consumption records, providing a data-driven foundation for our analysis. According to the MNO's policy, MU n consumes his short-term data first, i.e., $\hat{e}_t^n = (\bar{e}_t^n - x_t^n)^+$. When MU 's available short-term data is exhausted, the system automatically utilizes the long-term data allowance. Once both short-term and long-term data reserves are depleted, any additional data consumption incurs overage charges at a rate π of per unit. Therefore, after the MU's data consumption, the volume of left short-term data and long-term data would become

$$\begin{cases} \hat{e}_t^n = (\bar{e}_t^n - x_t^n)^+, \\ \hat{q}_t^n = \left(\bar{q}_t^n + (\bar{e}_t^n - x_t^n)^-\right)^+. \end{cases} \quad (5.3)$$

One-slot User Payoff MU's single-period payoff comprises three components: (1) utility from data consumption, (2) potential overage charges, and (3) net gains from data trading. Formally, MU n derives utility $u_n(x)$ from consuming x units of data, where $u_n(x)$ is an increasing and concave utility function capturing diminishing marginal returns. When demand exceeds the available quota (after accounting for trading and initial allocation), the MU incurs an overage charge $\pi(x_t^n - e_t^n - q_t^n - a_t^n)^+$. The trading decision a_t^n generates income $p_t^s \cdot (-a_t^n)^+$ when selling surplus data (i.e., $a_t^n < 0$) or a cost $p_t^b \cdot (a_t^n)^+$ when purchasing additional data (i.e., $a_t^n > 0$). Therefore, the *one-slot payoff* of MU n is

$$v_t^n\left(e_t^n, q_t^n, \boldsymbol{p}_t, a_t^n, x_t^n\right) = \\ u_n(x_t^n) - \pi\left(x_t^n - e_t^n - q_t^n - a_t^n\right)^+ + p_t^s \cdot (-a_t^n)^+ - p_t^b \cdot (a_t^n)^+. \quad (5.4)$$

To account for the randomness in data usage x_t^n, we derive the one-slot expected payoff of MU n by taking the expectation over x_t^n as follows:

$$\bar{v}_t^n\left(e_t^n, q_t^n, \boldsymbol{p}_t, a_t^n\right) = \int_0^{+\infty} v_t^n\left(e_t^n, q_t^n, \boldsymbol{p}_t, a_t^n, x\right) f_n(x) dx. \quad (5.5)$$

To illustrate the key insights of the optimal trading policy, we take $\bar{z}_t^n$ as the MU n's decision variable of data trading, which is equivalent to a_t^n. Hence we rewrite the MU's *one-slot expected payoff* as follows:

$$\bar{v}_t^n\left(e_t^n, q_t^n, \boldsymbol{p}_t, \bar{z}_t^n\right) = W(\bar{z}_t^n) + J(e_t^n + q_t^n - \bar{z}_t^n, \boldsymbol{p}_t), \quad (5.6)$$

where $W_n(\bar{z})$ and $J(z, \boldsymbol{p}_t)$ are given by

$$W(\bar{z}) \triangleq \int_0^{+\infty} \left[u_n(x) - \pi(x - \bar{z})^+\right] f_n(x) dx, \quad (5.7)$$

$$J(z, \boldsymbol{p}_t) \triangleq p_t^s \cdot (z)^+ - p_t^b \cdot (-z)^+. \quad (5.8)$$

Multi-slot Data Trading and Rollover Building upon MU's one-slot expected payoff defined in (5.6), we now extend our analysis to a multi-slot optimization framework. To establish this temporal linkage, we first characterize the state

5.1 Market Model

transition dynamics between consecutive time slots, which consists of two distinct cases:

- For any time slot t that does not coincide with a monthly boundary, the data volumes evolve as follows:

$$\begin{cases} e_{t+1}^n = \hat{e}_t^n, \\ q_{t+1}^n = \hat{q}_t^n. \end{cases} \tag{5.9}$$

- At the monthly boundary where time slot t corresponds to the end of a billing cycle, the short-term data $\hat{e}_t^n$ expires, while the long-term data $\hat{q}_t^n$ will be carried over to the next month and also added to the short-term data quota of the next month. Moreover, the monthly data allowance is fully renewed. Therefore, the next-slot data volume is given by

$$\begin{cases} e_{t+1}^n = \hat{q}_t^n, \\ q_{t+1}^n = Q^n. \end{cases} \tag{5.10}$$

Let $V_t^n(e_t^n, q_t^n, \boldsymbol{p}_t)$ represent the value function for MU n at time t, defined as the supremum of expected total discounted future payoffs from slot t until contract termination, conditioned on the current data allocations (e_t^n, q_t^n) and prevailing market prices $\boldsymbol{p}t = (pt^b, p_t^s)$. Accordingly, we model the multi-slot data trading process of MNO n as the following dynamic programming.

Problem 5.1 Consider the k-th time slot of the m-th month, i.e., $t = K(m-1)+k$. The MU's value function is defined by the following three cases:

1. If $m = M$ and $k = K$, the value function is

$$V_t^n(e_t^n, q_t^n, \boldsymbol{p}_t) = \max_{\bar{z}_t^n \geq 0} \{J(e_t^n + q_t^n - \bar{z}_t^n, \boldsymbol{p}_t) + W(\bar{z}_t^n)\}. \tag{5.11}$$

2. If $m < M$ and $k = K$, the value function is

$$\begin{aligned} V_t^n(e_t^n, q_t^n, \boldsymbol{p}_t) = \max_{\bar{z}_n \geq 0} \{ & J(e_t^n + q_t^n - \bar{z}_t^n, \boldsymbol{p}_t) + W(\bar{z}_t^n) \\ & + \delta \cdot \mathbb{E}_t \left[V_{t+1}^n (\hat{q}_t^n, Q^n, \boldsymbol{p}_{t+1}) \right] \}. \end{aligned} \tag{5.12}$$

3. If $k < K$, the value function is

$$\begin{aligned} V_t^n(e_t^n, q_t^n, \boldsymbol{p}_t) = \max_{\bar{z}_t^n \geq 0} \{ & J(e_t^n + q_t^n - \bar{z}_t^n, \boldsymbol{p}_t) + W(\bar{z}_t^n) \\ & + \delta \cdot \mathbb{E}_t \left[V_{t+1}^n (\hat{e}_t^n, \hat{q}_t^n, \boldsymbol{p}_{t+1}) \right] \}. \end{aligned} \tag{5.13}$$

Case 1 corresponds to the contract-ending slot. On the right-hand-side (RHS), the terms inside the brackets represent the one-slot expected payoff.

Case 2 corresponds to the ending slot in each month (excluding the contract-ending month). We use $\mathbb{E}_t[\cdot]$ to denote $\mathbb{E}_{x_n}\left[\mathbb{E}_{\boldsymbol{p}_{t+1}}[\cdot]\right]$ for brevity. The term $\delta \cdot \mathbb{E}_t\left[V_{t+1}^n\left(\hat{q}_t^n, Q^n, \boldsymbol{p}_{t+1}\right)\right]$ represents the expected maximal discounted payoff from slot $t+1$ to the end of the contract. The parameter $\delta \in (0, 1)$ is the time discount.

Case 3 corresponds to the other time slots. Similarly, we have substituted (5.9) in the third term of the RHS of (5.13).

In each time slot t, MU n needs to make his optimal data trading decision $\bar{z}_t^{n*}$ based on the trading prices $\boldsymbol{p}_t$ and his leftover data volume (e_t^n, q_t^n), while taking into account his random data demand x_t^n.

5.1.3 MNO's Decision

We formulate the MNO's pricing problem by building upon the established user model, where each MU n determines the optimal trading decision a_t^n based on residual data allocations and the prevailing trading prices $\boldsymbol{p}_t = \{p_t^s, p_t^b\}$. Depending on their supply-demand balance, MUs may act as sellers, buyers, or abstain from trading. Consequently, the aggregate market demand—summed across all purchasing MUs—is given by

$$D_t(\boldsymbol{p}_t) = \sum_{n \in \mathcal{N}} \left(a_t^n\right)^+, \tag{5.14}$$

and the aggregate market supply—summed across all selling MUs—is given by

$$S_t(\boldsymbol{p}_t) = \sum_{n \in \mathcal{N}} \left(-a_t^n\right)^+. \tag{5.15}$$

As shown in (5.14) and (5.15), the trading decision of MUs depends on the price vector $\boldsymbol{p}_t$. The total demand $D_t(\boldsymbol{p}_t)$ and the total supply $S_t(\boldsymbol{p}_t)$ lead to the total transaction quantity $\min\{D_t(\boldsymbol{p}_t), S_t(\boldsymbol{p}_t)\}$. Moreover, the MNO benefits $p_t^b - p_t^s$ from each unit of transaction data. Based on the above discussion, we formulate the MNO's revenue-maximizing pricing problem as follows:

Problem 5.2 (MNO's Pricing Problem)

$$\max_{\boldsymbol{p}_t \geq 0} \left(p_t^b - p_t^s\right) \cdot \min\left\{S_t(\boldsymbol{p}_t), D_t(\boldsymbol{p}_t)\right\}. \tag{5.16}$$

5.2 MU's Trading Policy

We analyze the MU's optimal data trading policy under two distinct scenarios:

- Plain Trading: The MU determines their trading strategy to maximize the total discounted payoff for the current month, assuming no data rollover occurs at the month's end.
- Rollover-Involved Trading: The MU optimizes their trading decisions to maximize the long-term discounted payoff, accounting for data rollover at the end of the current month.

If the MNO does not offer rollover, MUs will always engage in plain trading. If the MNO does offer rollover, MUs will follow plain trading in the final month of the contract period (since no further rollover applies), and rollover-involved trading in all prior months (where unused data can be carried forward). Next, we examine the trading policies under both scenarios and highlight their key differences. Our analysis is applicable to a generic MU, thus we omit the superscript n for MU unless explicitly required for clarity.

5.2.1 Plain Trading

Now we study the plain trading case. Consider the slot $t = K(M-1) + k$ and $k \in \{1, 2, \ldots, K\}$, given the leftover data (e, q) and the trading prices $\boldsymbol{p} = \{p^s, p^b\}$, the optimal trading policy corresponds to a pair of thresholds $\{L_{k,M}^{\text{Plain}}(p^b), U_{k,M}^{\text{Plain}}(p^s)\}$. Moreover, the corresponding optimal trading action is

$$\bar{z}_t^* = \begin{cases} L_{k,M}^{\text{Plain}}(p^b), & \text{if } e + q < L_{k,M}^{\text{Plain}}(p^b), \\ e + q, & \text{if } L_{k,M}^{\text{Plain}}(p^b) \leq e + q \leq U_{k,M}^{\text{Plain}}(p^s), \\ U_{k,M}^{\text{Plain}}(p^s), & \text{if } e + q > U_{k,M}^{\text{Plain}}(p^s). \end{cases} \quad (5.17)$$

The above discussion shows that the optimal plain trading policy is a *target interval policy* specified by the buy-up-to threshold $L_{k,M}^{\text{Plain}}(p^b)$ and the sell-down-to threshold $U_{k,M}^{\text{Plain}}(p^s)$. More specifically, if the MU's total leftover data volume $e + q$ is less than $L_{k,M}^{\text{Plain}}(p^b)$, then the MU will buy extra data such that the leftover data volume reaches $L_{k,M}^{\text{Plain}}(p^b)$. In contrast, when the leftover data $e+q$ is more than $U_{k,M}^{\text{Plain}}(p^s)$, then the MU should sell some data such that the leftover data volume reaches $U_{k,M}^{\text{Plain}}(p^s)$. Finally, if the leftover data $e + q$ is within the above thresholds, then MU does not need to trade data.

Figure 5.1a depicts the optimal plain trading policy. The horizontal and vertical axes represent the MU's short-term and long-term data, respectively. Hence each point of this plane corresponds to a specific data volume (e, q). The blue region

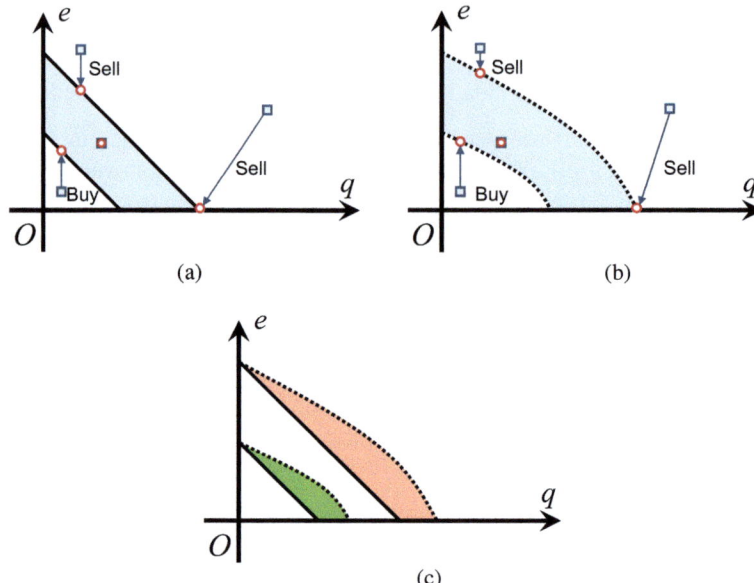

Fig. 5.1 Illustrations of the optimal trading policy. (**a**) Plain trading policy. (**b**) Rollover-involved policy. (**c**) Comparison

in Fig. 5.1a is generated by the *selling line* and the *buying line*, i.e., $L_{k,M}^{\text{Plain}}(p^b) \le e + q \le U_{k,M}^{\text{Plain}}(p^s)$. The states in the blue region represent those states where the MU do not need to trade data. The blue squares in Fig. 5.1a denote the MU's data volume before trading, i.e., (e_t, q_t). The red circles in Fig. 5.1a correspond to the MU's data volume after trading, i.e., $(\bar{e}_t, \bar{q}_t)$. The blue arrows illustrate the MU's optimal trading action.

5.2.2 Rollover-involved Trading

Now we study the rollover-involved trading. Similarly, we find that the optimal trading policy also has a target-interval structure. Let $\{L_{k,m}^{\text{Roll}}(p^b, q), U_{k,m}^{\text{Roll}}(p^s, q)\}$ denote the pair of thresholds, and the corresponding optimal trading action is

$$\bar{z}_t^* = \begin{cases} L_{k,m}^{\text{Roll}}(p^b, q), & \text{if } e + q < L_{k,m}^{\text{Roll}}(p^b, q), \\ e + q, & \text{if } L_{k,m}^{\text{Roll}}(p^b, q) \le e + q \le U_{k,m}^{\text{Roll}}(p^s, q), \\ U_{k,m}^{\text{Roll}}(p^s, q), & \text{if } e + q > U_{k,m}^{\text{Roll}}(p^s, q). \end{cases} \quad (5.18)$$

5.2 MU's Trading Policy

The analysis above demonstrates that the optimal trading policy involving rollover remains a target interval policy, consistent with the basic trading scenario. However, unlike the plain trading case, the buy-up-to threshold and the sell-down-to threshold depend not only on the trading price but also on the remaining long-term data quota. This additional dependence arises because unused long-term data carries over and influences costs in the subsequent month.

Figure 5.1b depicts the trading policy incorporating rollover for the k-th time slot in the m-th month (where $m < M$). The blue region bounded by the selling curve and buying curve indicates states where no trading occurs. A comparison between Fig. 5.1b and a reveals that the rollover mechanism transforms the linear selling and buying thresholds (seen in Fig. 5.1a) into nonlinear curves. In the following, we further analyze the trading thresholds and explore the implications of the rollover mechanism.

5.2.3 Impact of Rollover Mechanism

Building on the optimal trading policy analysis, we next examine how the rollover mechanism influences MUs trading decisions. For the same k-th time slot across different months under fixed trading prices, the associated thresholds adhere to the following conditions:

$$\begin{cases} L_{k,1}^{\text{Roll}}(p^b, q) \geq L_{k,2}^{\text{Roll}}(p^b, q) \geq \ldots \geq L_{k,M}^{\text{Plain}}(p^b), \\ U_{k,1}^{\text{Roll}}(p^s, q) \geq U_{k,2}^{\text{Roll}}(p^s, q) \geq \ldots \geq U_{k,M}^{\text{Plain}}(p^s). \end{cases} \quad (5.19)$$

We illustrate the above results by combining Fig. 5.1a and b together in Fig. 5.1c. Our analysis reveals two distinct operational regions (Region I and Region II), each corresponding to different optimal trading strategies:

- Region I: The optimal policy involves purchasing data when guided by the dashed curves (rollover-involved trading), whereas no trading occurs under the solid lines (plain trading). This indicates that the rollover mechanism incentivizes MUs to acquire more data compared to scenarios without rollover.
- Region II: The optimal policy transitions from data selling (under plain trading, represented by solid lines) to no-trading (under rollover conditions, shown as dashed curves). This shift demonstrates that the rollover mechanism diminishes MUs' propensity to sell data, consequently reducing market liquidity relative to conventional trading scenarios.

Our analysis reveals that the rollover mechanism induces MUs to maintain higher data inventories by shifting both the buying and selling curves upward. This strategic behavior yields two key advantages:

- Risk mitigation: Reduces exposure to overage charges.
- Revenue potential: Enables opportunistic selling when prices are more favorable.

Having established MUs' optimal trading behavior, we now turn to examine the MNO's pricing strategy optimization.

5.3 MNO's Optimal Pricing

This chapter investigates the MNO's revenue optimization problem while accounting for MUs' optimal data trading behavior. Restoring the superscript n to denote individual MUs, we formulate the total trading market demand at time slot t as follows: Given each MU's residual data volumes and prevailing market prices $p_t = \{p_t^s, p_t^b\}$, the aggregate demand from buyers is expressed as:

$$D_t(p_t^b) = \sum_{n \in \mathcal{N}} \left(L_t^n(p_t^b, q_t^n) - e_t^n - q_t^n \right)^+, \tag{5.20}$$

where $L_t^n(p_t^b, q_t^n)$ denotes the buy-up-to threshold. Similarly, we express the total trading market supply (from sellers) as follows:

$$S_t(p_t^s) = \sum_{n \in \mathcal{N}} \left(e_t^n + q_t^n - U_t^n(p_t^s, q_t^n) \right)^+, \tag{5.21}$$

where $U_t^n(p_t^s, q_t^n)$ is the sell-down-to threshold of MU n in time slot t.

Before we present the MNO's optimal pricing, let us first introduce the concept of *perfect competition*. In economics, perfect competition is defined by several conditions (see [40] for more details). One of these conditions is that *no participant has the market power to set prices*. In our problem, the data trading market is not perfectly competitive, since the MNO determines the prices. In addition, it has been theoretically demonstrated that a perfectly competitive market will reach an equilibrium where the quantity supplied equals the quantity demanded [40]. Next we show the somewhat surprising result that the MNO's revenue-maximizing pricing still clears the market, even though the data trading market does not exhibit the perfect competition.

Figure 5.2 presents three distinct market equilibrium scenarios. Each sub-figure employs a standardized coordinate system where the vertical axis represents price p and the horizontal axis represents quantity. The intersecting demand $D_t(p)$ and supply $S_t(p)$ curves characterize each market scenario, with their equilibrium points demonstrating the price-quantity dynamics under different conditions. Figure 5.2 presents three distinct market equilibrium scenarios. Each sub-figure employs a standardized coordinate system where the vertical axis represents price p and the horizontal axis represents quantity. The intersecting demand $D_t(p)$ and supply $S_t(p)$ curves characterize each market scenario, with their equilibrium points demonstrating the price-quantity dynamics under different conditions.

5.3 MNO's Optimal Pricing

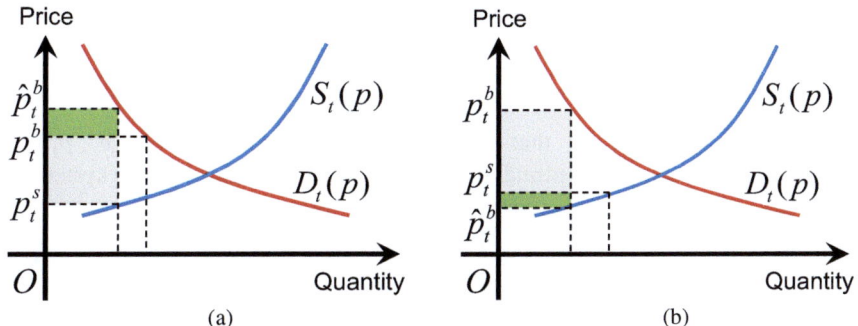

Fig. 5.2 An illustration of the market clear pricing. (**a**) Demand > supply. (**b**) Demand < supply

- Figure 5.2a: When market demand exceeds supply (i.e., $S_t(p_t^s) < D_t(p_t^b)$), the MNO's baseline revenue equals the gray-shaded area $(p_t^b - p_t^s) \cdot S_t(p_t^s)$. The MNO can strategically increase revenue by adjusting the buying price from p_t^b to $\hat{p}_t^b$. This price intervention generates additional revenue represented by the green-shaded area, demonstrating the marginal revenue gain from price optimization.
- Figure 5.2b: When market supply exceeds demand (i.e., $S_t(p_t^s) > D_t(p_t^b)$), the MNO can optimize revenue by strategically lowering the selling price from p_t^s to $\hat{p}_t^s$. This price adjustment yields a revenue increment represented by the green-shaded area.

Building on these insights, we conclude that the MNO's optimal pricing strategy should equilibrate market demand and supply for data trading. We formally characterize the MNO's optimal pricing scheme in Theorem 5.1. To establish clear notation, we define the inverse demand and supply functions respectively as:

$$\begin{aligned} P_{t,D}(\theta) &= D_t^{-1}(\theta), \\ P_{t,S}(\theta) &= S_t^{-1}(\theta), \end{aligned} \tag{5.22}$$

where $\theta \geq 0$ represents the quantity of total market demand for $P_{t,D}(\cdot)$ or supply for $P_{t,S}(\cdot)$.

Theorem 5.1 *In time slot t, the MNO's revenue-maximizing prices, denoted by $\{\tilde{p}_t^s, \tilde{p}_t^b\}$, are given by*

$$\begin{cases} \tilde{p}_t^s = P_{t,S}(\theta^*), \\ \tilde{p}_t^b = P_{t,D}(\theta^*), \end{cases} \tag{5.23}$$

where θ^* denotes the optimal transaction quantity satisfying

$$P_{t,D}(\theta^*) + \theta^* \cdot P'_{t,D}(\theta^*) = P_{t,S}(\theta^*) + \theta^* \cdot P'_{t,S}(\theta^*). \tag{5.24}$$

Theorem 5.1 establishes that the revenue-maximizing prices $(\tilde{p}_t^s, \tilde{p}_t^b)$ are uniquely determined by the optimal transaction quantity θ^* through the system:

$$\begin{aligned} \tilde{p}_t^s &= P_{t,S}(*), \\ \tilde{p}_t^b &= P_{t,D}(*), \end{aligned} \tag{5.25}$$

where $P_{t,S}()$ and $P_{t,D}()$ represent the inverse supply and demand functions respectively. For practical implementation, the MNO can estimate the demand and supply curves through the data-driven approach based on MU profiles.

5.4 Summary

This chapter presents a comprehensive economic analysis of data trading markets that integrate rollover mechanisms, establishing a novel framework to understand the dynamic interaction between two critical flexibility dimensions: temporal flexibility (enabled by data rollover) and user-driven flexibility (facilitated by peer-to-peer trading platforms). Through rigorous theoretical modeling and numerical simulations, we have systematically evaluated the market viability and welfare implications of this hybrid system, with particular attention to how these complementary flexibilities create multiplicative rather than merely additive value.

Our research yields three fundamental insights with significant theoretical and practical implications: First, the temporal flexibility introduced by rollover mechanisms serves as a powerful market catalyst for user-driven trading systems. By allowing unused data allowances to persist across billing cycles, rollover creates a more liquid and stable secondary market. Second, we identify a unique positive feedback loop between these flexibility dimensions: rollover mechanisms increase the tradable data supply in secondary markets, which in turn makes rollover allowances more valuable by providing additional usage pathways. Third, the dual-flexibility system exhibits remarkable market resilience. Even under conditions of demand fluctuation or network congestion, the interplay between rollover buffers and trading opportunities maintains system stability. These findings collectively demonstrate that temporal and user-identity flexibility are not merely compatible features, but rather mutually reinforcing components of an advanced data service ecosystem.

Chapter 6
Interplay between Time and User-identity Flexibility

Abstract This chapter focuses on the integrated time-user flexibility and further investigate how bounded rational MUs behave to it. This chapter will unveil the impact of MU bounded rationality on the economic benefit of flexible data services.

Keywords Consumption mechanism · Trading policy

6.1 Market Model

We consider a telecommunication market where an MNO provides wireless data service to the mobile users (MUs). The MNO offers a monthly contract characterized by a specific supply pattern and consumption mechanism. The contract, denoted as $\{Q, \Pi\}$, involves the subscriber paying a monthly subscription fee Π, for data usage up to the cap Q. For instance, CMHK's pricing structure includes a subscription fee of 40 for a data cap of 3GB.

Supply Pattern The supply pattern is linked to the trading service, characterized by the price pair $\pi = \{\pi_b, \pi_s\}$. Specifically, MUs have the option to sell leftover data within their monthly data cap Q to other MUs or the MNO at the unit price π_s, and purchase additional data from others or the MNO at the unit price π_b. The trading price pair π leads to four different supply patterns, denoted by $\eta \in \{\text{No}, \text{Sl}, \text{BY}, \text{Td}\}$.

- The case of $\eta = \text{No}$ corresponds to $\pi_s = 0$ and $\pi_b = \infty$, which represents that the MNO does not offer trading service. Hence the MUs merely subscribe to a fixed monthly contract $\{Q, \Pi\}$.
- The case of $\eta = \text{Sl}$ corresponds to $\pi_s > 0$ and $\pi_b = \infty$, which represents that the MNO only allows MUs to sell the leftover data (within the data cap Q) back to the MNO.
- The case of $\eta = \text{BY}$ corresponds to $\pi_s = 0$ and $\pi_b < \infty$, which represents that the MNO only allows the MUs to buy extra data from the MNO.
- The case of $\eta = \text{Td}$ corresponds to $\pi_s > 0$ and $\pi_b < \infty$, which represents that the MNO allows the MUs to trade (sell and buy) data.

© The Author(s), under exclusive license to Springer Nature Singapore Pte Ltd. 2026
Z. Wang et al., *Mobile Data Services*, SpringerBriefs in Computer Science,
https://doi.org/10.1007/978-981-95-1343-7_6

Consumption Mechanism The consumption mechanism, denoted by $\kappa \in \{T, R\}$, corresponds to the adoption of the rolling service, which determines when the leftover data expires.

- The case of $\kappa = T$ corresponds to the traditional mechanism without consumption flexibility. The leftover data cap expires at the end of each month.
- The case of $\kappa = R$ corresponds to the rollover mechanism with one-month valid period. It allows the leftover data cap in the previous month to be consumed in the current month and expires in the next month.

We formulate the economic interactions between the MNO and the MUs as a two-stage game. In Stage I, the MNO first determines the consumption mechanism $\kappa \in \{T, R\}$ and the supply pattern $\eta \in \{\text{No}, \text{Sl}, \text{BY}, \text{Td}\}$, then it further decides the optimal pricing strategy, including the subscription price Π, the selling price π_s (if $\eta \in \{\text{Sl}, \text{Td}\}$), and the buying price π_b (if $\eta \in \{\text{BY}, \text{Td}\}$). In Stage II, each MU decides his subscription, data consumption, and trading action to maximize his expected monthly payoff.

6.1.1 User Model

MUs usually sign a one-year or two-year contract with the MNO. Hence we consider a slotted time horizon consisting of T months indexed by $t \in \{1, 2, \ldots, T\}$.

Effective Data Cap An MU's effective data cap refers to the total consumable data quota that the MU owns in a particular month. It depends on the consumption mechanism $\kappa \in \{T, R\}$. If the MNO adopts the traditional mechanism (i.e., $\kappa = T$), then the MU's effective data cap is always Q in each month t. If the MNO adopts the rollover mechanism (i.e., $\kappa = R$), then the MU can consume the unused data in the previous month, in addition to the current monthly data cap Q. We let r_t denote the rollover data from month $t-1$ to month t, which will expire at the end of month t. Therefore, the MU's effective data cap in month t is $Q + r_t$ under the rollover mechanism.

Data Demand We model an MU's data demand in month t as a random variable d_t, which takes integer values (i.e., $d_t \in \{0, 1, 2, \ldots\}$). Here the wireless data amount is measured in the data unit, i.e., 1KB or 1MB according to the MNO's billing practice. Let $\bar{d} = \mathbb{E}[d_t]$ denote the mean of the data demand, and let $f(\cdot)$ denote the probability mass function (PMF) of the data demand.

Data Consumption Decision We denote x_t as the MU's actual data consumption in month t, which is determined by the MU. Accordingly, we let $\boldsymbol{x} = \{x_t, 1 \leq t \leq T\}$ denote the data consumption vector over the T-month contract period. As we will see later, a payoff-maximizing MU may not consume data exactly up to his data demand d_t. For example, an MU may obtain a larger payoff by restricting his data demand and selling part of the consumable data. Furthermore, we let θ denote the

6.1 Market Model

MU's data valuation of unit data consumption, which is the MU's characteristic. Hence, the utility derived by the MU with type θ under the data consumption x_t is θx_t. Here we adopt a linear utility for ease of exposition and facilitation of analysis. It is commonly used in related literatures on data pricing (see, e.g., [38, 57, 60]).

Data Trading Decision We denote a_t as the MU's trading action in month t. Accordingly, we let $\boldsymbol{a} = \{a_t, 1 \leq t \leq T\}$ denote the data trading vector over the T-month contract period. Moreover, the sign of a_t corresponds to the MU's selling or buying choice. The case of $a_t = 0$ represents that no trading happens in month t. The case of $a_t > 0$ represents that the MU sells a_t units of data. The case of $a_t < 0$ represents that the MU buys $|a_t|$ units of data. Therefore, the MU's selling income and the buying cost is $\pi_s[a_t]^+$ and $\pi_b[-a_t]^+$, respectively. Here $[\cdot]^+ = \max(0, \cdot)$ denotes the projection onto the nonnegative orthant.

User's Payoff The MU's payoff in month t is defined as the difference between his utility of consuming data (i.e., θx_t) and the total payment (i.e., $\Pi + \pi_b[-a_t]^+ - \pi_s[a_t]^+$), as follows

$$U(\Pi, \boldsymbol{\pi}, \theta, x_t, a_t) \triangleq \theta x_t + \pi_s[a_t]^+ - \pi_b[-a_t]^+ - \Pi. \tag{6.1}$$

The MU's total payoff over the T-month contract period (with the consumption vector $\boldsymbol{x}$ and the trading vector $\boldsymbol{a}$) is given by

$$U_T(\Pi, \boldsymbol{\pi}, \theta, \boldsymbol{x}, \boldsymbol{a}) = \sum_{t=1}^{T} U(\Pi, \boldsymbol{\pi}, \theta, x_t, a_t). \tag{6.2}$$

Note that not every consumption vector $\boldsymbol{x}$ and trading vector $\boldsymbol{a}$ are feasible. Next we introduce two feasibility conditions, i.e., the Potential Demand Condition (PDC) and the Effective Cap Condition (ECC). Given the random data demand d_t in month t, a payoff-maximizing MU does not necessarily fulfill his entire data demand d_t. Hence, we have the following potential demand condition

$$0 \leq x_t \leq d_t, \quad \forall t \in \{1, 2, 3, \ldots, T\} \quad \text{(PDC)}. \tag{6.3}$$

In each month t, the total data that the MU consumes and sells cannot exceed the effective data cap that he has in this month. Recall that the effective data cap depends on the consumption mechanism $\kappa \in \{T, R\}$. Hence we have the following two sets of ECC.

- Under the traditional mechanism $\kappa = T$, the effective cap is always the same as the monthly data cap Q, thus we have

$$x_t + a_t \leq Q, \quad \forall t \in \{1, 2, \ldots, T\} \quad \text{(ECC - T)}. \tag{6.4}$$

- Under the rollover mechanism $\kappa = R$, the effective cap is $Q + r_t$. Here r_t is the rollover data in month t, which cannot be carried to the next month. Therefore, we have

$$\begin{cases} x_t + a_t \leq Q + r_t, & \forall t \in \{1, 2, \ldots, T\}, \\ r_{t+1} = [Q + r_t - x_t - a_t]_0^Q, & \forall t \in \{1, 2, \ldots, T-1\}, \\ r_1 = 0, \end{cases} \quad \text{(ECC - R)},$$

(6.5)

where $[s]_y^z = s - [s - z]^+ + [y - s]^+$ is the projection of s onto the interval $[y, z]$.

6.1.2 User's Consumption and Trading Problem

A rational MU tends to maximize his total payoff $U_T(\Pi, \boldsymbol{\pi}, \theta, \boldsymbol{x}, \boldsymbol{a})$ by jointly optimizing over the consumption vector $\boldsymbol{x}$ and trading vector $\boldsymbol{a}$. In practice, however, the MU has to make sequential decisions in each month. The event sequence in month t is as follows:

1. The MU observes his random data demand d_t and the available rollover data r_t (if any).
2. The MU decides his data consumption x_t and trading action a_t subject to PDC and ECC.

Note from the above sequence that the MU's decisions in different months are independent under the traditional mechanism $\kappa = \text{T}$, since there is no rollover data. But the MU's decisions are coupled under the rollover mechanism $\kappa = \text{R}$. In this case, the MU needs to make the consumption and trading decisions in the online context. That is, the MU tends to decide the data consumption and the trading action without the knowledge of the future data demand.

Under the traditional mechanism $\kappa = \text{T}$, maximizing the total payoff is equivalent to maximizing the single-month payoff, which leads to Problem 6.1.

Problem 6.1 *Under the traditional mechanism* T, *the MU determines his data consumption x_t and trading action a_t to maximize his monthly payoff, i.e.,*

$$\max \quad U(\Pi, \boldsymbol{\pi}, \theta, x_t, a_t) \tag{6.6a}$$

$$\text{s.t.} \quad 0 \leq x_t \leq d_t, \tag{6.6b}$$

$$x_t + a_t \leq Q, \tag{6.6c}$$

$$x_t \in \mathbb{Z}, \ a_t \in \mathbb{Z}, \tag{6.6d}$$

$$\text{var.} \quad x_t, \ a_t. \tag{6.6e}$$

6.1 Market Model

In Problem 6.1, the objective (6.6a) is the single-month payoff under the data consumption x_t and the trading action a_t. The inequalities (6.6b) and (6.6c) correspond to the PDC and ECC-T, respectively.

Under the rollover mechanism $\kappa = \text{R}$, the MU needs to make the data consumption and trading decisions in an online context. In this case, the MU needs a pair of consumption and trading policies that map the current state to the consumption and trading decisions. To facilitate our later analysis, we first formulate the *off-line problem* (assuming the data demand vector is known in advance) as the benchmark.

Problem 6.2 (Off-line Scenario Under $\kappa = \text{R}$) *If the MU knew the data demand $d = \{d_t, 1 \leq t \leq T\}$ in advance, then he determines the data consumption vector x and the data trading vector a to maximize his total payoff, i.e.,*

$$U_T^{Off}(Q, \Pi, \pi, \text{R}, \theta, d) = \max \quad U_T(\Pi, \pi, \theta, x, a) \tag{6.7a}$$

$$\text{s.t.} \quad 0 \leq x_t \leq d_t, \; \forall t \in \{1, 2, \ldots, T\} \tag{6.7b}$$

$$x_t + a_t \leq Q + r_t, \; \forall t \in \{1, 2, \ldots, T\} \tag{6.7c}$$

$$r_1 = 0, \tag{6.7d}$$

$$r_{t+1} = [Q + r_t - x_t - a_t]_0^Q, \; \forall t \geq 1 \tag{6.7e}$$

$$x_t \in \mathbb{Z}, \; a_t \in \mathbb{Z}, \; \forall t \in \{1, 2, \ldots, T\}, \tag{6.7f}$$

$$\text{var.} \quad x = \{x_t, \forall t\}, \; a = \{a_t, \forall t\}. \tag{6.7g}$$

In Problem 6.2, the objective (6.7a) is the MU's total payoff across the T-months contract period. Constraint (6.7b) represents the potential demand condition. Constraints (6.7c), (6.7d), and (6.7e) represent the effective cap conditions. Here we let $U_T^{Off}(Q, \Pi, \pi, \text{R}, \theta, d)$ denote the MU's optimal payoff in hindsight, given the data demand d.

In the *online context*, for example, in month t, the MU only observe his monthly data demand d_t and the rollover data r_t. Therefore, the MU needs a consumption policy $\mathfrak{X}(\cdot)$ and a trading policy $\mathfrak{A}(\cdot)$ that map from the current state (i.e., d_t and r_t) to the data consumption $x_t = \mathfrak{X}(d_t, r_t)$ and the trading action $a_t = \mathfrak{A}(d_t, r_t)$, respectively. Accordingly, we denote the MU's *online total payoff* in the T months under the consumption policy $\mathfrak{X}(\cdot)$ and the trading policy $\mathfrak{A}(\cdot)$ as follow:

$$U_T^{On}(Q, \Pi, \pi, \text{R}, \theta, d, \mathfrak{X}, \mathfrak{A}) = \sum_{t=1}^{T} U\big(\Pi, \pi, \theta, \mathfrak{X}(d_t, r_t), \mathfrak{A}(d_t, r_t)\big), \tag{6.8}$$

where $r_1 = 0$ and r_t updates according to

$$r_{t+1} = \big[Q + r_t - \mathfrak{X}(d_t, r_t) - \mathfrak{A}(d_t, r_t)\big]_0^Q, \; \forall t \in \{1, 2, \ldots, T-1\}. \tag{6.9}$$

Formally, we define the regret of the MU to be the expected payoff difference between the optimal off-line decisions and the online policies.

$$\text{Regret}_T(\mathfrak{X}, \mathfrak{A}) \triangleq U_T^{Off}(Q, \Pi, \pi, R, \theta, \boldsymbol{d}) - U_T^{On}(Q, \Pi, \pi, R, \theta, \boldsymbol{d}, \mathfrak{X}, \mathfrak{A}). \tag{6.10}$$

The MU's goal is to find the policies $\mathfrak{X}(\cdot)$ and $\mathfrak{A}(\cdot)$, and try to minimize the regret. We say that the consumption policy $\mathfrak{X}(\cdot)$ and the trading policy $\mathfrak{A}(\cdot)$ perform well if the corresponding regret is sub-linear as a function of T, i.e., $\text{Regret}_T(\mathfrak{X}, \mathfrak{A}) = o(T)$ [61]. The sub-linear regret implies that on average the policies perform as well as the optimal decisions in hindsight, i.e., $\lim_{T \to \infty} \frac{1}{T} \cdot \text{Regret}_T(\mathfrak{X}, \mathfrak{A}) = 0$. Hence we formulate the MU's online decision making process in Problem 6.3.

Problem 6.3 (Online Scenario Under $\kappa = R$) *Under rollover mechanism R, the MU determines his policies $\mathfrak{X}(\cdot)$ and $\mathfrak{A}(\cdot)$ to minimize the time-average regret:*

$$\min \quad \lim_{T \to \infty} \frac{\text{Regret}_T(\mathfrak{X}, \mathfrak{A})}{T} \tag{6.11a}$$

$$\text{s.t.} \quad 0 \leq \mathfrak{X}(d_t, r_t) \leq d_t, \tag{6.11b}$$

$$r_{t+1} = \left[Q + r_t - \mathfrak{X}(d_t, r_t) - \mathfrak{A}(d_t, r_t)\right]_0^Q, \; \forall\, t \in \{1, 2, \ldots, T-1\}, \tag{6.11c}$$

$$r_1 = 0, \tag{6.11d}$$

$$\text{var.} \quad \mathfrak{X}(\cdot), \; \mathfrak{A}(\cdot). \tag{6.11e}$$

6.1.3 Trading Completion Ratio

Now we have formulated the MU's data consumption and trading problems under the two consumption mechanisms ($\kappa = T$ and $\kappa = R$) in Problems 6.1 and 6.3. In Sect. 6.2, we will derive the MU's optimal data consumption and the optimal trading amount given the stochastic demand d and the rollover data r (if any), which is denoted by

$$\{x^*(d, r), a^*(d, r)\}. \tag{6.12}$$

Ideally, the MU can achieve the maximal payoff if he can consume and trade data according to the optimal strategy $\{x^*(d, r), a^*(d, r)\}$. In practice, however, we need to note that MUs may not be able to trade data based on the optimal trading amount $a^*(d, r)$ due to the MNO's trading service implementation. First, additional operations on trading will reduce the bounded rational MU's trading amount. Second, the minimal trading amount (determined by the MNO) reduces the MU's trading amount as well. To capture the difference between the optimal and actual trading amounts, we let **trading completion ratio** $\sigma \in [0, 1]$ denote the average proportion of the *actual trading amount* to the *optimal trading amount* in the market. Strictly speaking, the trading completion ratio (or simply completion ratio) varies for different MUs in different months. Here we consider the *average value* in the market and take it as our future work to study its dependence on the trading service. Mathematically, the MU's *actual trading amount* is

$$\sigma \cdot a^*(d, r). \tag{6.13}$$

In practice, the completion ratio σ is close to 1, if the trading process is simple to operate (e.g., automatically done like [7]) and the minimum trading amount is small enough (e.g., 1MB or even 1KB). The completion ratio also can be very small, if the trading process requires complicated operations or the minimum trading amount is very large (e.g., 1GB). In Sect. 6.3, we will elaborate how the completion ratio σ affects the interrelationship between the consumption flexibility and the supply flexibility. Before that, we first study MUs' decisions on the data consumption, data trading, and the subscription choice in Sect. 6.2.

6.2 User's Consumption and Trading Policy

We investigate MUs' optimal decisions in Stage II. Specifically, we first analyze the MU's optimal data consumption and trading decisions under different mechanisms, then derive the MU's expected payoff and characterize the subscription choice.

6.2.1 *Optimal Consumption and Trading Decisions*

We first derive the MUs' optimal decisions under traditional mechanism and rollover mechanism. We then introduce the impact of rollover data on MUs' optimal decisions.

Traditional Mechanism Theorem 6.1 presents the MU's optimal consumption and trading decisions under the traditional mechanism $\kappa = \text{T}$, i.e., the solution of Problem 6.1.

Theorem 6.1 *Under the traditional mechanism* T, *given the MU's data demand* d_t, *the optimal data consumption* $x_t^*(d_t)$ *and trading action* $a_t^*(d_t)$ *are given by*

$$x_t^*(d_t) = \begin{cases} 0, & \text{if } \theta < \pi_s, \\ \min\{d_t, Q\}, & \text{if } \pi_s \leq \theta \leq \pi_b, \\ d_t, & \text{if } \theta > \pi_b, \end{cases} \tag{6.14}$$

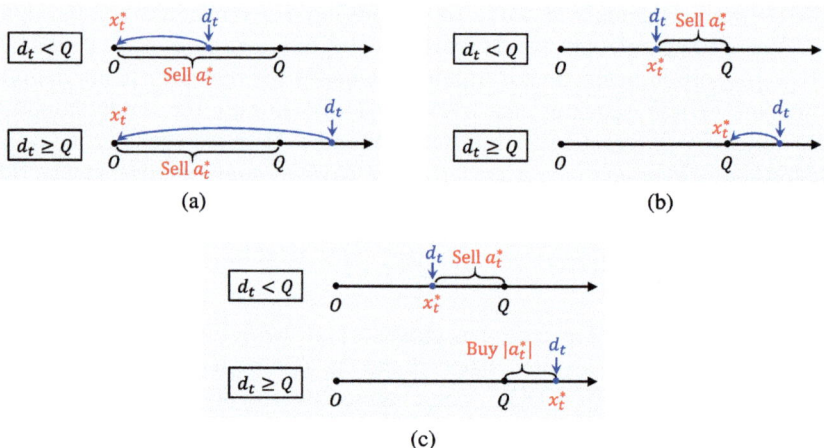

Fig. 6.1 Users' optimal data consumption x_t^* and trading action a_t^* under traditional mechanism. (**a**) Low valuation. (**b**) Medium valuation. (**c**) High valuation

and

$$a_t^*(d_t) = Q - x_t^*(d_t). \tag{6.15}$$

Theorem 6.1 shows the dependence between the MU's data valuation θ and the trading price pair $\pi = \{\pi_b, \pi_s\}$. First, the low-valuation MU (i.e., $\theta < \pi_s$) will not consume any data, but will sell his monthly data cap. As shown in Fig. 6.1a, we have $x_t^* = 0$ and $a_t^* = Q$ no matter how large the demand d_t is. Second, the medium-valuation MU (i.e., $\pi_s \leq \theta \leq \pi_b$) will fulfill his data demand within the data cap and sell the leftover data. That is, medium-valuation MU will not buy additional data to meet the data demand exceeding the data cap. As shown in Fig. 6.1b, if the data demand is less than the monthly data cap (i.e., $d_t < Q$), then we have $x_t^* = d_t$ and $a_t^* = Q - d_t > 0$. If the data demand exceeds the monthly data cap (i.e., $d_t \geq Q$), then we have $x_t^* = Q$ and $a_t^* = 0$. Third, the high-valuation MU (i.e., $\theta > \pi_b$) is willing to buy extra data to fulfill his entire demand (even it exceeds the monthly data cap) and sell his leftover data (when the data demand is less than the monthly data cap). As shown in Fig. 6.1c, we always have $x_t^* = d_t$ and $a_t^* = Q - d_t$.

Rollover Mechanism Theorem 6.2 presents the MU's optimal consumption and trading policies under the rollover mechanism R.

Theorem 6.2 *Under the rollover mechanism R, the following consumption policy $\mathfrak{X}^*(d_t, r_t)$ and trading policy $\mathfrak{A}^*(d_t, r_t)$ achieve a sub-linear regret in T,*

$$\mathfrak{X}^*(d_t, r_t) = \begin{cases} 0, & \text{if } \theta < \pi_s, \\ \min(d_t, Q + r_t), & \text{if } \pi_s \leq \theta \leq \pi_b, \\ d_t, & \text{if } \theta \geq \pi_b, \end{cases} \tag{6.16}$$

6.2 User's Consumption and Trading Policy

and

$$\mathfrak{A}^*(d_t, r_t) = [r_t - \mathfrak{X}^*(d_t, r_t)]^+ - [\mathfrak{X}^*(d_t, r_t) - Q - r_t]^+. \quad (6.17)$$

We have two remarks regarding the results in Theorem 6.2.

First, the trading policy $\mathfrak{A}^*(\cdot)$ is defined based on the consumption policy $\mathfrak{X}^*(\cdot)$. Given the consumption policy, the trading policy in (6.17) is the most *forward-looking* strategy in the sense that it fully takes the advantage of the consumption flexibility. More specifically, the trading policy in (6.17) can be expressed as follows:

$$\mathfrak{A}^*(d_t, r_t) = \begin{cases} r_t - \mathfrak{X}^*(d_t, r_t), & \text{if } \mathfrak{X}^*(d_t, r_t) < r_t, \\ 0, & \text{if } r_t \leq \mathfrak{X}^*(d_t, r_t) \leq Q + r_t, \\ Q + r_t - \mathfrak{X}^*(d_t, r_t), & \text{if } \mathfrak{X}^*(d_t, r_t) > Q + r_t, \end{cases} \quad (6.18)$$

which implies that given the data consumption $\mathfrak{X}^*(d_t, r_t)$, the MU should only sell the *leftover rollover data* (that expires at the end of the current month), but should not sell the *leftover monthly data cap* (that can roll over to the next month). By doing this, the MU does not waste any data and make full use of the rolling service to tackle his demand uncertainty in the future.

Second, the consumption policy $\mathfrak{X}^*(\cdot)$ is the most *myopic* strategy, which maximizes the current month payoff given the most *forward-looking* trading strategy. The low valuation MU (i.e., $\theta < \pi_s$) will not consume any data, i.e., $\mathfrak{X}(d_t, r_t) = 0$. As shown in Fig. 6.2a, we always have $x_t^* = 0$ and $a_t^* = r_t$. The medium valuation MU (i.e., $\pi_s \leq \theta \leq \pi_b$) will only fulfill his data demand within the *effective data cap*, i.e., $\mathfrak{X}^*(d_t, r_t) = \min(d_t, Q + r_t)$, instead of buying extra data to meet the data demand exceeding it. As shown in Fig. 6.2b, we have $x_t^* = \min(d_t, Q + r_t)$ and $a_t^* = [r_t - x_t^*]^+ \geq 0$. The high valuation MU (i.e., $\theta > \pi_b$) will fulfill his entire data demand, i.e., $\mathfrak{X}^*(d_t, r_t) = d_t$. As shown in Fig. 6.2c, we have $x_t^* = d_t$ for any d_t no matter how large the data demand d_t is.

Impact of Rollover on Consumption and Trading Policies Now we have derived the MU's optimal data consumption and trading decisions under two consumption mechanisms. By comparing Theorems 6.1 and 6.2, we obtain the following insights regarding the impact of data rollover:

- *Data Consumption:* The optimal consumption decisions in (6.14) and (6.16) differ only in the availability of the rollover data r_t. If $r_t = 0$, then we have $x_t^*(d_t) = \mathfrak{X}^*(d_t, 0)$. If $r_t > 0$, then we have $x_t^*(d_t) \leq \mathfrak{X}^*(d_t, r_t)$. Overall, rollover data helps satisfy MUs' data consumption demand.
- *Data Trading:* The optimal trading decisions in (6.15) and (6.17) reflect the value of the rolling choice. Specifically, the trading decision without rolling is to sell all the leftover data in the current month (otherwise it expires). While the trading decision with rolling only needs to sell the leftover data that expires in the current month. Overall, rollover mechanism allows MUs to hold more data to accommodate the future data demand fluctuation.

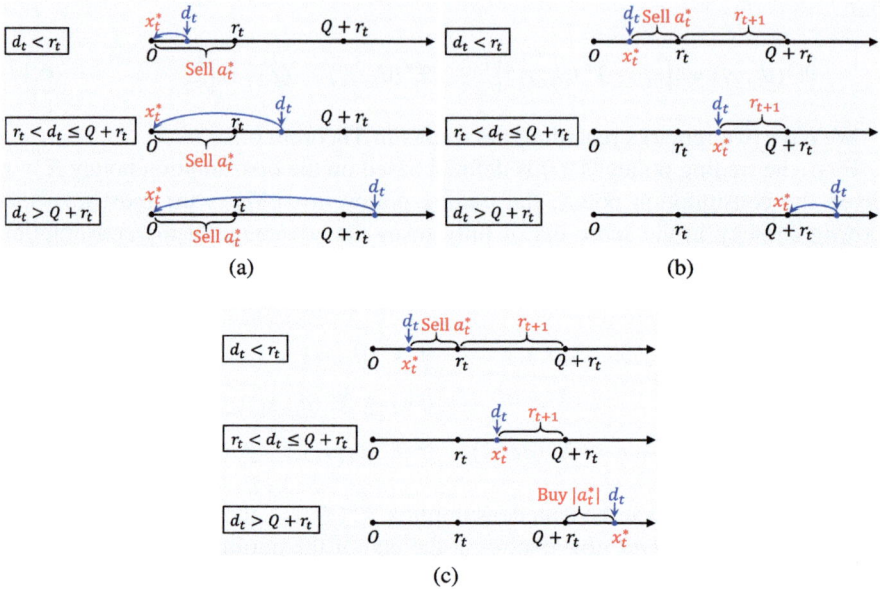

Fig. 6.2 Users' optimal data consumption x_t^* and trading action a_t^* under rollover mechanism. (**a**) Low valuation. (**b**) Medium valuation. (**c**) High valuation

6.2.2 Expected Monthly Payoff

Next we derive the MU's expected monthly payoff under the aforementioned optimal data consumption and trading decisions. We will focus on the MU's *expected monthly payoff*, and neglect the month index t for notation simplicity.

Traditional Mechanism We derive the MU's expected monthly payoff under the traditional mechanism $\kappa = \mathrm{T}$ based on the optimal decisions in Theorem 6.1. Specifically, we substitute the MU's optimal data consumption and trading action in Theorem 6.1 into (6.1), and obtain the MU's monthly payoff as follows

$$U(Q, \Pi, \boldsymbol{\pi}, \mathrm{T}, \theta, d) = \theta \cdot x^*(d) + \pi_s \left[a^*(d)\right]^+ - \pi_b \left[-a^*(d)\right]^+ - \Pi, \quad (6.19)$$

where the data demand d is a random variable that changes in different months. We take the expectation over the data demand d and derive the MU's expected monthly payoff as follows

$$\bar{U}(Q, \Pi, \boldsymbol{\pi}, \mathrm{T}, \theta) = \mathbb{E}_d\left[U(Q, \Pi, \boldsymbol{\pi}, \mathrm{T}, \theta, d)\right]. \quad (6.20)$$

Next we explain the detailed expression of (6.20) for different types of MUs.

6.2 User's Consumption and Trading Policy

- The high valuation MU ($\theta > \pi_b$) is willing to buy extra data to fulfill the entire data demand (in the heavy-demand month) or sell the leftover data (in the light-demand month). Hence the expected monthly utility of consuming data is $\mathbb{E}_d\{\theta d\} = \theta \bar{d}$. The expected monthly selling amount is $\mathbb{E}_d\{[Q-d]^+\}$ and the expected buying amount is $\mathbb{E}_d\{[d-Q]^+\}$.
- The medium valuation MU ($\pi_s \leq \theta \leq \pi_b$) will only fulfill his data demand within the data cap or sell the leftover data (in the light-demand month). Hence the expected utility of consuming data is $\mathbb{E}_d\{\theta(d-[d-Q]^+)\}$ and the expected selling amount is $\mathbb{E}_d\{[Q-d]^+\}$.
- The low valuation MU ($\theta < \pi_s$) will not consume data but sell the entire data cap, which leads to a zero utility and the constant selling amount Q.

For notation simplicity, we define the expected monthly selling amount $\phi_s(Q, T)$ and the expected monthly buying amount $\phi_b(Q, T)$ as follows:

$$\text{Expected Selling Amount}: \quad \phi_s(Q, T) \triangleq \mathbb{E}_d\{[Q-d]^+\} \quad (6.21a)$$

$$= \sum_{d=0}^{D}[Q-d]^+ f(d), \quad (6.21b)$$

$$\text{Expected Buying Amount}: \quad \phi_b(Q, T) \triangleq \mathbb{E}_d\{[d-Q]^+\} \quad (6.21c)$$

$$= \sum_{d=0}^{D}[d-Q]^+ f(d). \quad (6.21d)$$

Accordingly, the MU's expected monthly payoff (6.20) can be expressed as follows:

$$\bar{U}(Q, \Pi, \pi, T, \theta) = \begin{cases} \theta \bar{d} + \pi_s \phi_s(Q, T) - \pi_b \phi_b(Q, T) - \Pi, & \text{if } \theta > \pi_b, \\ \theta\left[\bar{d} - \phi_b(Q, T)\right] + \pi_s \phi_s(Q, T) - \Pi, & \text{if } \pi_s \leq \theta \leq \pi_b, \\ \pi_s Q - \Pi, & \text{if } \theta < \pi_s. \end{cases} \quad (6.22)$$

Rollover Mechanism We derive the MU's expected monthly payoff under the rollover mechanism $\kappa = R$ based on the optimal consumption and trading policies in Theorem 6.2. Given the data demand d and the rollover data r in the current month, we substitute the consumption and trading policies in Theorem 6.2 into (6.1), and express the MU's payoff as follows

$$U(Q, \Pi, \pi, R, \theta, d, r) = \theta \cdot \mathfrak{X}^*(d, r) + \pi_s\left[\mathfrak{A}^*(d, r)\right]^+ - \pi_b\left[-\mathfrak{A}^*(d, r)\right]^+ - \Pi, \quad (6.23)$$

where the data demand d and the rollover data r are two random variables that change in different months. Specifically, the data demand d in each month follows the distribution $f(d)$. As we will show later on, the rollover data r in different months exhibits a Markov property. For notation simplicity, we denote $p_\theta(r, \hat{r})$ as the transition probability from r to $\hat{r}$ for the MUs with data valuation θ, and let $\bar{p}_\theta(\cdot)$ denote the corresponding stationary distribution of the Markov chain. Next we first explain the transition of the rollover data in Proposition 6.1, then derive the transition probability $p_\theta(r, \hat{r})$ and the corresponding stationary distribution $\bar{p}_\theta(\cdot)$.

Proposition 6.1 *For rollover mechanism, under the consumption and trading policies in Theorem 6.2, the rollover data in the next month is given by*

$$\hat{r} = \begin{cases} Q, & \text{if } \theta < \pi_s, \\ [Q + r - d]_0^Q, & \text{if } \theta \geq \pi_s, \end{cases} \tag{6.24}$$

where d and r represent the data demand and the rollover data in the current month, respectively.

Proposition 6.1 summarizes the transition of the rollover data for MUs with different data valuation θ, which leads to the following observations.

- For the low-valuation MUs (i.e., $\theta < \pi_s$), the optimal data rolling over to the next month does not depend on the data demand d or the rollover data r in the current month.
- For the medium-valuation MUs (i.e., $\pi_s \leq \theta \leq \pi_b$) and the high-valuation MUs (i.e., $\theta > \pi_b$), the data rolling decisions are the same, which depend on the data demand d and the rollover data r in the current month. This implies the Markov property.

To evaluate the expected monthly payoff under rollover mechanism, we leverage Proposition 6.1 to derive the transition probability $p_\theta(r, \hat{r})$ and the stationary distribution $\bar{p}_\theta(\cdot)$ in Propositions 6.2 and 6.3.

Proposition 6.2 *For low-valuation MUs (i.e., $\theta < \pi_s$), the transition probability is*

$$p_\theta(r, \hat{r}) = \begin{cases} 1, & \text{if } \hat{r} = Q, \\ 0, & \text{otherwise,} \end{cases} \tag{6.25}$$

and the corresponding stationary distribution of the rollover data is

$$\bar{p}_\theta(r) = \begin{cases} 1, & \text{if } r = Q, \\ 0, & \text{otherwise.} \end{cases} \tag{6.26}$$

Proposition 6.2 shows that the rollover data is Q with the probability one in the stationary distribution for the low valuation MUs ($\theta < \pi_s$). This is because Proposition 6.1 indicates that the low-valuation MUs will not consume any data, but sell the data that cannot roll over in each month.

6.2 User's Consumption and Trading Policy

Proposition 6.3 *For the medium-valuation MUs (i.e., $\pi_s \leq \theta \leq \pi_b$) and the high-valuation MUs (i.e., $\theta > \pi_b$), the transition probability is*

$$p_\theta(r, \hat{r}) = \begin{cases} \sum_{d=0}^{r} f(d), & \text{if } \hat{r} = Q, \\ f(Q + r - \hat{r}), & \text{if } \hat{r} \in \{1, 2, \ldots, Q-1\}, \\ \sum_{d=Q+r}^{+\infty} f(d), & \text{if } \hat{r} = 0. \end{cases} \quad (6.27)$$

The corresponding stationary distribution of the rollover data satisfies

$$\begin{cases} \bar{p}_\theta(r) = \sum_{\tau=0}^{Q} \bar{p}_\theta(\tau) \cdot p_\theta(\tau, r), & \forall r \in \{0, 1, \ldots, Q\}, \\ 1 = \sum_{\tau=0}^{Q} \bar{p}_\theta(\tau). \end{cases} \quad (6.28)$$

For the medium-valuation MUs and high-valuation MUs, Proposition 6.3 shows that the stationary distribution of the rollover data depends on the MU's data demand $f(\cdot)$. Due to the complexity of the transition probability, we are not able to express the stationary distribution $\bar{p}_\theta(\cdot)$ in a closed form. Nevertheless, it does not affect our later analysis on the impact of rollover mechanism.

So far, we have introduced the stationary distribution $\bar{p}_\theta(\cdot)$ of the rollover data for different MUs. Although MUs usually have a finite subscription period, we will evaluate the expected monthly payoff based on the stationary distribution of rollover data. Such an evaluation captures an MU's long-term average benefit, which will affect its subscription behavior. Mathematically, the MU's expected monthly payoff under rollover mechanism is

$$\bar{U}(Q, \Pi, \boldsymbol{\pi}, R, \theta) = \mathbb{E}_{d,r}[U(Q, \Pi, \boldsymbol{\pi}, R, \theta, d, r)]$$
$$= \sum_{r=0}^{Q} \sum_{d=0}^{+\infty} U(Q, \Pi, \boldsymbol{\pi}, R, \theta, d, r) f(d) \bar{p}_\theta(r). \quad (6.29)$$

Next we explain the detailed expression of (6.29) for different types of MUs.

- The high valuation MU (i.e., $\theta > \pi_b$) is willing to buy extra data to fulfill his entire data demand (even if it exceeds the effective data cap) and sell the leftover rollover data. Hence the expected utility of consuming data is $\mathbb{E}\{\theta d\} = \theta \bar{d}$. The expected selling amount and the buying amount are $\mathbb{E}\{[r-d]^+\}$ and $\mathbb{E}\{[d-Q-r]^+\}$, respectively. Here the expectation is taken over the data demand d and the rollover data r (in the stationary state).

- The medium valuation MU (i.e., $\pi_s \le \theta \le \pi_b$) will only fulfill his data demand within the effective data cap and sell the leftover rollover data. Hence the expected utility of consuming data is $\mathbb{E}\{\theta(d - [d - Q - r]^+)\}$. The expected selling amount is $\mathbb{E}\{[r - d]^+\}$, which has the same expression as that for high-valuation MUs. Similarly, the expectation is taken over the data demand d and the rollover data r (in the stationary state).
- The low-valuation MU (i.e., $\theta < \pi_s$) will not consume data but sell the leftover rollover data. Hence the low-valuation MU has a zero utility, the expected monthly selling amount is Q.

For notation simplicity, we define the expected selling amount $\phi_s(Q, R)$ and the expected buying amount $\phi_b(Q, R)$ under rollover mechanism as follows

$$\phi_s(Q, R) \triangleq \mathbb{E}_{d,r}\{[r - d]^+\} = \sum_{r=0}^{Q}\sum_{d=0}^{D}[r - d]^+ f(d)\bar{p}_\theta(r), \qquad (6.30a)$$

$$\phi_b(Q, R) \triangleq \mathbb{E}_{d,r}\{[d - Q - r]^+\} = \sum_{r=0}^{Q}\sum_{d=0}^{D}[d - Q - r]^+ f(d)\bar{p}_\theta(r). \qquad (6.30b)$$

Accordingly, the MU's expected monthly payoff under the rollover mechanism is

$$\bar{U}(Q, \Pi, \pi, R, \theta) = \begin{cases} \theta\bar{d} + \pi_s\phi_s(Q, R) - \pi_b\phi_b(Q, R) - \Pi, & \text{if } \theta > \pi_b, \\ \theta\left[\bar{d} - \phi_b(Q, R)\right] + \pi_s\phi_s(Q, R) - \Pi, & \text{if } \pi_s \le \theta \le \pi_b, \\ \pi_s Q - \Pi, & \text{if } \theta < \pi_s. \end{cases} \qquad (6.31)$$

Unification and Comparison Based on the above analysis for traditional and rollover mechanisms, now we can derive a unified expression for the MU's expected monthly payoff under the consumption mechanisms $\kappa \in \{T, R\}$, as follows

$$\bar{U}(Q, \Pi, \pi, \kappa, \theta) = \begin{cases} \theta\bar{d} + \pi_s\phi_s(Q, \kappa) - \pi_b\phi_b(Q, \kappa) - \Pi, & \text{if } \theta > \pi_b, \\ \theta\left[\bar{d} - \phi_b(Q, \kappa)\right] + \pi_s\phi_s(Q, \kappa) - \Pi, & \text{if } \pi_s \le \theta \le \pi_b, \\ \pi_s Q - \Pi, & \text{if } \theta < \pi_s, \end{cases} \qquad (6.32)$$

where $\phi_s(Q, \kappa)$ is defined in (6.21a) and (6.30a). Moreover, $\phi_b(Q, \kappa)$ is defined in (6.21c) and (6.30b). Theorem 6.3 presents some properties of $\phi_b(Q, \kappa)$ and $\phi_s(Q, \kappa)$.

Theorem 6.3 *For arbitrary data demand distribution $f(d)$, the following properties are true.*

1. *The $\phi_s(Q, \kappa)$ is increasing and convex in Q. Moreover, $\phi_s(0, \kappa) = 0$ for any $\kappa \in \{T, R\}$.*

6.2 User's Consumption and Trading Policy

2. The $\phi_b(Q, \kappa)$ is decreasing and convex in Q. Moreover, $\phi_b(0, \kappa) = \bar{d}$ for any $\kappa \in \{T, R\}$.
3. For any $Q > 0$, we have

$$\phi_s(Q, R) < \phi_s(Q, T), \qquad (6.33a)$$

$$\phi_b(Q, R) < \phi_b(Q, T). \qquad (6.33b)$$

4. The following equation holds:

$$Q + \phi_b(Q, \kappa) - \phi_s(Q, \kappa) = \bar{d}. \qquad (6.34)$$

Theorem 6.3 explains how the data cap Q and the consumption mechanism κ affect the expected selling and buying amount at the presence of the random data demand. Figure 6.3 provides an illustration. We have the following remarks.

- Properties 1 and 2 in Theorem 6.3 show that as the data cap Q increases, the unit data cap increment leads to a larger selling amount increment and a smaller buying amount decrement. As shown in Fig. 6.3, the solid curves represent the expected selling amount $\phi_s(Q, \kappa)$, which are convexly increasing in the data cap Q. The dash curves represent the expected buying amount $\phi_b(Q, \kappa)$, which are convexly decreasing in the data cap Q.
- Property 3 indicates that the rollover mechanism R makes MUs more *patient* in the sense that it reduces the selling amount and the buying amount, as presented in (6.33). In Fig. 6.3, the two solid curves (or the two dash curves, respectively) verify the inequality $\phi_s(Q, R) < \phi_s(Q, T)$ (or $\phi_b(Q, R) < \phi_b(Q, T)$, respectively).
- Property 4 indicates the balance between the data demand and the data supply for any consumption mechanism. That is, the data supply $Q + \phi_b(Q, \kappa) - \phi_s(Q, \kappa)$ equals to the data demand $\bar{d}$. Particularity, if the data cap Q equals to the average data demand $\bar{d}$, then (6.34) implies that $\phi_b(\bar{d}, \kappa) = \phi_s(\bar{d}, \kappa)$, as shown by the black line in Fig. 6.3.

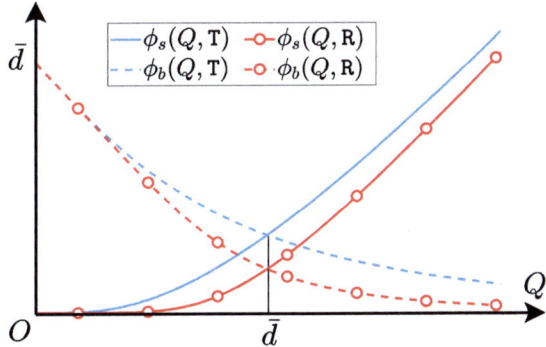

Fig. 6.3 An illustration of $\phi_b(Q, \kappa)$ and $\phi_s(Q, \kappa)$ for $\kappa \in \{T, R\}$

Recall that (6.32) denotes the MU's expected monthly payoff with the optimal consumption and trading decisions. It is the theoretically optimal case for the MU. To facilitate our later analysis, we take into account the trading completion ratio $\sigma \in [0, 1]$ and express the MU's *actual expected payoff* as follows:

$$\bar{U}(Q, \Pi, \boldsymbol{\pi}, \kappa, \sigma, \theta) =
\begin{cases}
\theta\left[\bar{d} - (1-\sigma)\phi_b(Q,\kappa)\right] + \sigma\pi_s\phi_s(Q,\kappa) - \sigma\pi_b\phi_b(Q,\kappa) - \Pi, & \theta > \pi_b, \\
\theta\left[\bar{d} - \phi_b(Q,\kappa)\right] + \sigma\pi_s\phi_s(Q,\kappa) - \Pi, & \pi_s \leq \theta \leq \pi_b, \\
\sigma\pi_s Q - \Pi, & \theta < \pi_s.
\end{cases}$$
(6.35)

The MU's actual expected payoff (6.35) will degenerate into (6.32) when the completion ratio $\sigma = 1$. As mentioned in Sect. 6.1.3, the trading completion ratio σ results from the MU's bounded rationality and the MNO's trading service implementation (i.e., additional operations and the minimal trading amount). In other words, the MU cannot adjust his decisions to compensate the reduction resulted from σ. Consequently, the MU cannot achieve the theoretically optimal payoff in practice. It is easy to show that the actual expected payoff $\bar{U}(Q, \Pi, \boldsymbol{\pi}, \kappa, \sigma, \theta)$ increases in the completion ratio σ.

6.2.3 User Subscription

Next we analyze MUs' subscription to MNOs, and derive the market partition outcome. In general, we suppose that an MU will subscribe to the MNO if and only if he can achieve a non-negative expected payoff. Theorem 6.4 summarizes the MUs' subscription decisions given the MNO's pricing outcome. Before introducing the main results in Theorem 6.4, we first discuss some conditions for the trading price pair $\boldsymbol{\pi}$ and the subscription fee Π, since some pricing outcomes are not sensible for a rational MNO. For example, $\pi_b < \pi_s$ would lead to arbitrage opportunities, and $\pi_s > \Pi/Q$ would induce MUs to sell the entire data cap. To avoid these trivial cases, we suppose that $\{\pi_b, \pi_s, \Pi\} \in \Psi(Q)$, where $\Psi(Q)$ is the feasible pricing set, defined as

$$\Psi(Q) \triangleq \left\{(\pi_b, \pi_s, \Pi), \ \Pi \geq 0, \ 0 \leq \pi_s \leq \frac{\Pi}{Q}, \ \pi_s \leq \pi_b\right\}.$$
(6.36)

6.3 MNO's Optimal Pricing

To facilitate our later discussion in Theorem 6.4, we split the feasible pricing set $\Psi(Q)$ in terms of the subscription fee, and define

$$\Psi_1(Q,\kappa) \triangleq \left\{(\pi_b, \pi_s, \Pi) \in \Psi(Q) : \Pi < \left[\bar{d} - \phi_b(Q,\kappa)\right]\pi_b + \sigma\pi_s\phi_s(Q,\kappa)\right\}, \tag{6.37a}$$

$$\Psi_2(Q,\kappa) \triangleq \left\{(\pi_b, \pi_s, \Pi) \in \Psi(Q) : \Pi \geq \left[\bar{d} - \phi_b(Q,\kappa)\right]\pi_b + \sigma\pi_s\phi_s(Q,\kappa)\right\}. \tag{6.37b}$$

Note that we have $\Psi_1(Q,\kappa) \bigcup \Psi_2(Q,\kappa) = \Psi(Q)$ and $\Psi_1(Q,\kappa) \bigcap \Psi_2(Q,\kappa) = \emptyset$. Intuitively, $\Psi_1(Q,\kappa)$ corresponds to a low subscription fee compared to the trading prices and the data cap. In contrast, $\Psi_2(Q,\kappa)$ corresponds to a high subscription fee. Both the two sets will be used to derive the threshold data valuation in Theorem 6.4.

Theorem 6.4 *Given the monthly contract $\{Q, \Pi\}$, the trading price pair $\boldsymbol{\pi}$, and the consumption mechanism κ, the MU with the data valuation θ will subscribe to the MNO if and only if*

$$\theta \geq \Theta(Q, \Pi, \boldsymbol{\pi}, \kappa), \tag{6.38}$$

where $\Theta(Q, \Pi, \boldsymbol{\pi}, \kappa)$ is the threshold data valuation, given by

$$\Theta(Q, \Pi, \boldsymbol{\pi}, \kappa) = \begin{cases} \dfrac{\Pi - \pi_s Q + (1-\sigma)\pi_s\phi_s(Q,\kappa)}{\bar{d} - \phi_b(Q,\kappa)} + \pi_s, & \text{if } (\pi_b, \pi_s, \Pi) \in \Psi_1(Q,\kappa), \\ \dfrac{\Pi - \sigma\pi_s\phi_s(Q,\kappa) + \sigma\pi_b\phi_b(Q,\kappa)}{\bar{d} - (1-\sigma)\phi_b(Q,\kappa)}, & \text{if } (\pi_b, \pi_s, \Pi) \in \Psi_2(Q,\kappa). \end{cases} \tag{6.39}$$

According to Theorem 6.4, the case of $\{\boldsymbol{\pi}, \Pi\} \in \Psi_1(Q,\kappa)$ corresponds to a low subscription fee. The subscribers includes part of the medium valuation MUs (i.e., $\Theta(Q, \Pi, \boldsymbol{\pi}, \kappa) < \theta < \pi_b$) and all of the high valuation MUs $\theta > \pi_b$. In contrast, the case of $\{\boldsymbol{\pi}, \Pi\} \in \Psi_2(Q,\kappa)$ corresponds to a high subscription fee. The subscribers are all high valuation MUs (i.e., $\theta > \Theta(Q, \Pi, \boldsymbol{\pi}, \kappa) > \pi_b$). Based on the aforementioned MU subscription decision, we will derive the MNO's revenue and analyze the MNO's pricing problem.

6.3 MNO's Optimal Pricing

In Stage I, the MNO decides the consumption mechanism, the supply pattern, and the pricing strategy. Next we first derive the MNO's revenue, and introduce the MNO's pricing problem. We then present the optimal pricing with and without selling service. Finally, we summarize the key insights.

6.3.1 MNO's Revenue

The MNO's revenue consists of the subscription fee and the income (or cost) from the trading service. Based on the MU's expected monthly payoff (6.35), we know that the MNO's expected monthly revenue from an MU with data valuation θ is

$$\bar{R}(Q, \Pi, \pi, \kappa, \sigma, \theta) = \begin{cases} \Pi - \sigma\pi_s\phi_s(Q,\kappa) + \sigma\pi_b\phi_b(Q,\kappa), & \text{if } \theta > \pi_b, \\ \Pi - \sigma\pi_s\phi_s(Q,\kappa), & \text{if } \pi_s \le \theta \le \pi_b, \\ \Pi - \sigma\pi_s Q, & \text{if } \theta < \pi_s, \end{cases} \quad (6.40)$$

where $\sigma \in (0, 1)$ is the average trading completion ratio in the market.

We consider MUs' heterogeneity in their data valuations (e.g., [29, 60]). More specifically, let $h(\cdot)$ and $H(\cdot)$ denote the PDF and the CDF of the data valuation, respectively. Based on the MUs' subscription choices in Sect. 6.2.3, the MNO's total revenue from the subscribers is

$$\tilde{R}(Q, \Pi, \pi, \kappa, \sigma) = \int_{\Theta(Q,\Pi,\pi,\kappa)}^{\theta_{\max}} \bar{R}(Q, \Pi, \pi, \kappa, \sigma, \theta) h(\theta) d\theta, \quad (6.41)$$

where $\Theta(Q, \Pi, \pi, \kappa)$ is the threshold data valuation given in (6.39). Similarly, the total payoff of the subscribers is given by

$$\tilde{U}(Q, \Pi, \pi, \kappa, \sigma) = \int_{\Theta(Q,\Pi,\pi,\kappa)}^{\theta_{\max}} \bar{U}(Q, \Pi, \pi, \kappa, \sigma, \theta) h(\theta) d\theta. \quad (6.42)$$

Note that (6.41) and (6.42) are the general formulations given the consumption flexibility $\kappa \in \{T, R\}$ and the supply pattern $\eta \in \{\text{No, Sl, BY, Td}\}$. The MNO's pricing problem has four different forms depending on the supply pattern $\eta \in \{\text{No, Sl, BY, Td}\}$. Next we introduce these cases one by one.

6.3.2 MNO's Pricing Problem

The MNO's pricing problem has four different forms depending on the supply pattern $\eta \in \{\text{No, Sl, BY, Td}\}$. Table 6.1 summarizes the difference of MNO's pricing problems under different supply patterns. The difference lies in the number of pricing variables.

For case $\eta = \text{Td}$, the MNO tends to maximize its total revenue $\tilde{R}(Q, \Pi, \pi_b, \pi_s, \kappa, \sigma)$ by deciding the optimal subscription Π, the selling price π_s, and the buying price π_b, as in Problem 6.4

6.3 MNO's Optimal Pricing

Table 6.1 Structure of the analysis of Stage I

Pricing variables		Supply			
		None (η = No)	Sell (η = S1)	Buy (η = BY)	Trade (η = Td)
Consumption	No (κ = T)	Π	(π_s, Π)	(π_b, Π)	(π_s, π_b, Π)
	Yes (κ = R)				

Problem 6.4 (MNO Pricing Problem η = Td)

$$\tilde{R}^\star_{\text{Td}}(Q, \kappa, \sigma) \triangleq \max_{\{\pi, \Pi\} \in \Phi(Q)} \tilde{R}(Q, \Pi, \pi_b, \pi_s, \kappa, \sigma). \tag{6.43}$$

For case $\eta = \text{BY}$, the MNO tends to fix the selling price $\pi_s = 0$ and maximize its total revenue $\tilde{R}(Q, \Pi, \pi_b, 0, \kappa, \sigma)$ via $\{\pi_b, \Pi\}$. For space limitation, we omit the mathematical formulation and let $\tilde{R}^\star_{\text{BY}}(Q, \kappa, \sigma)$ denote the optimal revenue of this case.

For case $\eta = \text{S1}$, the MNO tends to fix the buying price $\pi_b = +\infty$ and maximize its total revenue $\tilde{R}(Q, \Pi, \infty, \pi_s, \kappa, \sigma)$ via $\{\pi_s, \Pi\}$. We let $\tilde{R}^\star_{\text{S1}}(Q, \kappa, \sigma)$ denote the optimal revenue of this case.

For case $\eta = \text{No}$, the MNO tends to fix the trading price $(\pi_s, \pi_b) = (0, +\infty)$ and maximize its total revenue $\tilde{R}(Q, \Pi, \infty, 0, \kappa, \sigma)$ via the subscription fee Π. Similarly, we let $\tilde{R}^\star_{\text{No}}(Q, \kappa, \sigma)$ denote the optimal revenue of this case.

Next we investigate the MNO's pricing problems under the four supply patterns.

6.3.3 Pricing Without Buying Service

For presentation convenience, Sect. 6.3.3 presents the main results of supply patterns without buying service, i.e., $\eta \in \{\text{No}, \text{S1}\}$. We will proceed in two steps. Section 6.3.3.1 introduces the dominant advantage of consumption mechanism. Section 6.3.3.2 presents the optimal pricing.

6.3.3.1 Dominant Advantage of Consumption Mechanism:

Proposition 6.4 first introduces the MNO's dominant advantage in terms of adopting rollover mechanism.

Proposition 6.4 *Under the supply pattern $\eta \in \{\text{No}, \text{S1}\}$, for any contract $\{Q, \Pi\}$, the rollover mechanism R can increase the MNO's total revenue and MUs' total payoff, i.e.,*

$$\begin{aligned} \tilde{R}_\eta(Q, \Pi, \text{R}, \sigma) &\geq \tilde{R}_\eta(Q, \Pi, \text{T}, \sigma), \\ \tilde{U}_\eta(Q, \Pi, \text{R}, \sigma) &\geq \tilde{U}_\eta(Q, \Pi, \text{T}, \sigma), \end{aligned} \quad \forall \eta \in \{\text{No}, \text{S1}\}. \tag{6.44}$$

The equality holds if and only if $Q = 0$.

Proposition 6.4 implies that the rollover mechanism $\kappa = $ R leads to a win-win situation (for both the MNO and MUs) under an arbitrary monthly contract $\{Q, \Pi\}$ (not necessary the MNO's revenue-maximizing one) given the supply pattern without buying service. **From the MUs' aspect**, the rollover mechanism $\kappa = $ R definitely increases each MU's payoff under the same data cap and subscription fee, hence increases the total payoff of the entire user market. It also induces more MUs to subscribe to the MNO. **From the MNO's aspect**, the intuition is as follows:

- For supply pattern $\eta = $ No, the MNO's revenue from each MU is the monthly subscription fee. Rollover mechanism induces more MUs to subscribe to the MNO, thus also increases the MNO's total revenue.
- For supply pattern $\eta = $ S1, the MNO's revenue is the total subscription fee minus MUs' selling benefits. On the one hand, rollover mechanism increases each MU's payoff, thus more MUs tend to subscribe to the MNO. On the other hand, the inequality $\phi_s(Q, R) < \phi_s(Q, T)$ in Theorem 6.3 implies that the rollover mechanism reduces the expected selling amount. Therefore, the rollover mechanism brings the MNO a larger revenue.

6.3.3.2 Optimal Pricing Without Buying Service:

Theorems 6.5 and 6.6 introduce the optimal pricing under supply pattern $\eta = $ No and $\eta = $ S1, respectively. For notation simplicity, we define a critical parameter θ^* as follows

$$\theta^* = \frac{1 - H(\theta^*)}{h(\theta^*)}, \tag{6.45}$$

where $h(\cdot)$ and $H(\cdot)$ are the PDF and the CDF of the data valuation distribution, respectively. Moreover, θ^* is unique for any data valuation distribution $h(\cdot)$ satisfying the increasing failure rate (IFR). Note that the IFR condition is widely adopted in the nonlinear pricing literatures (see, e.g., the foundational work of [62]).

Theorem 6.5 *Under the supply pattern $\eta = $ No, for any consumption mechanism $\kappa \in \{T, R\}$, the MNO's optimal subscription fee is given by*

$$\Pi^* = \theta^* \left[\bar{d} - \phi_b(Q, \kappa) \right]. \tag{6.46}$$

Moreover, the MNO's total revenue and MUs' total payoff under the optimal price Π^* are

$$\tilde{R}^\star_{\mathrm{No}}(Q, \kappa, \sigma) = \left[\bar{d} - \phi_b(Q, \kappa) \right] \theta^* \left[1 - H\left(\theta^*\right) \right], \tag{6.47a}$$

$$\tilde{U}^\star_{\mathrm{No}}(Q, \kappa, \sigma) = \left[\bar{d} - \phi_b(Q, \kappa) \right] \theta^* \left[\int_{\theta^*}^{\theta_{max}} \theta h(\theta) \mathrm{d}\theta - \theta^* + \theta^* \cdot H\left(\theta^*\right) \right]. \tag{6.47b}$$

6.3 MNO's Optimal Pricing

To understand the insights in Theorem 6.5, let us recall the inequality $\phi_b(Q, R) < \phi_b(Q, T)$ in Theorem 6.3. Mathematically, this inequality leads to $\bar{d} - \phi_b(Q, R) > \bar{d} - \phi_b(Q, T)$. Given this inequality, (6.46) implies that the MNO tends to charge a higher subscription fee for rollover mechanism R, since the MUs can enjoy the consumption flexibility. Moreover, (6.47) implies that the rollover mechanism increases both the total revenue of MNO and the total payoff of MUs under the optimal subscription fee. This result is also consistent in Proposition 6.4.

Theorem 6.6 *Under the supply pattern $\eta = S1$, for any consumption mechanism $\kappa \in \{T, R\}$, the MNO's optimal subscription fee Π^* and the optimal selling price π_s^* satisfy*

$$\begin{cases} \pi_s^* \geq 0, \\ \Pi^* = \sigma \pi_s^* \phi_s(Q, \kappa) + \theta^* \left[\bar{d} - \phi_b(Q, \kappa)\right]. \end{cases} \quad (6.48)$$

Moreover, the MNO's total revenue and MUs' total payoff under the optimal prices are

$$\tilde{R}_{S1}^{\star}(Q, \kappa, \sigma) = \left[\bar{d} - \phi_b(Q, \kappa)\right] \theta^* \left[1 - H\left(\theta^*\right)\right], \quad (6.49a)$$

$$\tilde{U}_{S1}^{\star}(Q, \kappa, \sigma) = \left[\bar{d} - \phi_b(Q, \kappa)\right] \theta^* \left[\int_{\theta^*}^{\theta_{max}} \theta h(\theta) d\theta - \theta^* + \theta^* H\left(\theta^*\right)\right]. \quad (6.49b)$$

Theorem 6.6 has twofold implications. First, comparing (6.49) and (6.47) shows that selling service cannot increase MNO's revenue or MUs' payoff. Second, comparing (6.48) and (6.46) shows a *substitute* trade-off between the subscription fee and the selling price, i.e., a higher selling price leads to a higher subscription fee. That is, the MNO can recover its revenue loss (due to a higher selling price) through a higher subscription fee.

6.3.4 Pricing with Buying Service

Section 6.3.4 presents the main results of supply patterns with buying service, i.e., $\eta \in \{BY, Td\}$. Recall that Proposition 6.4 has shown that the consumption flexibility (offered by rollover mechanism) exhibits a dominant advantage under the supply pattern without buying service, i.e., $\eta \in \{No, S1\}$. In the presence of buying service (i.e., $\eta \in \{BY, Td\}$), however, such a dominant advantage does not exist anymore. This means that the absence of buying service makes the consumption flexibility stand out intensively in the sense that the adoption of rollover mechanism benefits both MNO and MUs for any monthly contract $\{Q, \Pi\}$. When the supply pattern includes buying service, we investigate the optimal pricing for $\eta = BY$ and $\eta = Td$ in Theorems 6.7 and 6.8, respectively.

Theorem 6.7 *Under the supply pattern $\eta = \text{BY}$ and the consumption mechanism κ, the MNO's optimal subscription fee Π^* and the optimal buying price π_b^* satisfy*

$$\begin{cases} 0 \leq \pi_b^* \leq \theta^*, \\ \Pi^* = [\bar{d} - (1-\sigma)\phi_b(Q, \kappa)]\theta^* - \sigma \pi_b^* \phi_b(Q, \kappa). \end{cases} \quad (6.50)$$

Moreover, the MNO's total revenue and MUs' total payoff under the optimal prices are

$$\tilde{R}_{\text{BY}}^\star(Q, \kappa, \sigma) = [\bar{d} - (1-\sigma)\phi_b(Q, \kappa)]\theta^* [1 - H(\theta^*)], \quad (6.51a)$$

$$\tilde{U}_{\text{BY}}^\star(Q, \kappa, \sigma) = [\bar{d} - (1-\sigma)\phi_b(Q, \kappa)]\theta^* \left[\int_{\theta^*}^{\theta_{max}} \theta h(\theta) d\theta - \theta^* + \theta^* H(\theta^*)\right]. \quad (6.51b)$$

We elaborate Theorem 6.7 from two aspects. First, (6.50) in Theorem 6.7 implies a *complementary* trade-off between the subscription fee and the buying price, i.e., a higher buying price corresponds to a lower subscription fee. That is, the MNO can balance its revenue from MUs' subscription fee and MUs' extra data purchase. Second, (6.51) together with the inequality $\phi_b(Q, \text{R}) < \phi_b(Q, \text{T})$ in Theorem 6.3 shows that the rollover mechanism $\kappa = \text{R}$ increases both the MNO's total revenue and MUs' total payoff, i.e.,

$$\begin{aligned} \tilde{R}_{\text{BY}}^\star(Q, \text{T}, \sigma) &\leq \tilde{R}_{\text{BY}}^\star(Q, \text{R}, \sigma), \\ \tilde{U}_{\text{BY}}^\star(Q, \text{T}, \sigma) &\leq \tilde{U}_{\text{BY}}^\star(Q, \text{R}, \sigma), \end{aligned} \quad (6.52)$$

where the equality holds if and only if $\sigma = 1$. Recall that σ measures the average proportion of the *actual trading amount* to the *theoretically optimal trading amount* in the market. Therefore, the inequalities in (6.52) suggest that the rollover mechanism $\kappa = \text{R}$ can enhance market welfare, provided that not all MUs are capable of adopting the theoretically optimal buying strategy. This situation arises when the minimal trading amount is not extremely small, or when MUs do not fully exhibit rational behavior.

Theorem 6.8 *Under the supply pattern $\eta = \text{Td}$ and the consumption mechanism κ, the optimal subscription fee Π^* and the optimal trading price pair π^* satisfy*

$$\begin{cases} 0 \leq \pi_b^* \leq \theta^*, \\ 0 \leq \pi_s^* \leq \dfrac{[Q - \phi_s(Q, \kappa)]\theta^*}{Q - \sigma \phi_s(Q, \kappa)}, \\ \Pi^* + [\pi_b^* \phi_b(Q, \kappa) - \pi_s^* \phi_s(Q, \kappa)]\sigma = [\bar{d} - (1-\sigma)\phi_b(Q, \kappa)]\theta^*. \end{cases} \quad (6.53)$$

6.3 MNO's Optimal Pricing

Moreover, the MNO's total revenue and MUs' total payoff under the optimal prices are

$$\tilde{R}^\star_{\text{Td}}(Q,\kappa,\sigma) = \left[\bar{d} - (1-\sigma)\phi_b(Q,\kappa)\right]\theta^*\left[1 - H(\theta^*)\right], \tag{6.54a}$$

$$\tilde{U}^\star_{\text{Td}}(Q,\kappa,\sigma) = \left[\bar{d} - (1-\sigma)\phi_b(Q,\kappa)\right]\left[\int_{\theta^*}^{\theta_{max}} \theta \cdot h(\theta)d\theta - \theta^* + \theta^* H(\theta^*)\right]. \tag{6.54b}$$

The pricing policy (6.53) in Theorem 6.8 introduces a three-way trade-off among the optimal subscription fee and the optimal trading price pair. We have two observations:

- It is possible that $\pi_s^* = 0$ according to (6.53). This means that removing selling service will not harm the MNO's total revenue if the other pricing variables are proper.
- We always have $\pi_b^* < \infty$ according to (6.53). This indicates the necessity of offering buying service for the revenue-maximizing MNO.

6.3.5 Summary

Based on Theorems 6.5, 6.6, 6.7, and 6.8, we are now ready to summarize the key insights regarding the consumption flexibility and the supply flexibility.

First, the inequality $\phi_b(Q, \text{R}) < \phi_b(Q, \text{T})$ in Theorem 6.3 indicates that for the supply pattern $\eta \in \{\text{No}, \text{Sl}\}$, we have

$$\begin{aligned}\text{MNO Revenue:} \quad & \tilde{R}^\star_\eta(Q,\text{T},\sigma) < \tilde{R}^\star_\eta(Q,\text{R},\sigma), \\ \text{User Payoff:} \quad & \tilde{U}^\star_\eta(Q,\text{T},\sigma) < \tilde{U}^\star_\eta(Q,\text{R},\sigma),\end{aligned} \quad \forall \sigma \in [0,1]. \tag{6.55}$$

While under the supply pattern $\eta \in \{\text{BY}, \text{Td}\}$, we have

$$\begin{aligned}\tilde{R}^\star_\eta(Q,\text{T},\sigma) &\leq \tilde{R}^\star_\eta(Q,\text{R},\sigma), \\ \tilde{U}^\star_\eta(Q,\text{T},\sigma) &\leq \tilde{U}^\star_\eta(Q,\text{R},\sigma),\end{aligned} \tag{6.56}$$

where the equality holds if and only if $\sigma = 1$. The two observations lead to Insight 6.1.

Insight 6.1 *The impact of consumption flexibility is as follows:*

- *Given a supply pattern without buying service (i.e., $\eta \in \{\text{No}, \text{Sl}\}$), adopting consumption flexibility is always win-win for the MNO and MUs.*

- *Given a supply pattern with buying service (i.e., $\eta \in \{\text{BY}, \text{Td}\}$), adopting consumption flexibility can further increase the welfare when not all MUs can trade data according to the theoretically optimal strategy.*

Second, comparing Eqs. (6.54), (6.51), (6.49), and (6.47) yields

$$\tilde{R}^{\star}_{\text{No}}(Q, \kappa, \sigma) = \tilde{R}^{\star}_{\text{S1}}(Q, \kappa, \sigma) < \tilde{R}^{\star}_{\text{BY}}(Q, \kappa, \sigma) = \tilde{R}^{\star}_{\text{Td}}(Q, \kappa, \sigma),$$
$$\tilde{U}^{\star}_{\text{No}}(Q, \kappa, \sigma) = \tilde{U}^{\star}_{\text{S1}}(Q, \kappa, \sigma) < \tilde{U}^{\star}_{\text{BY}}(Q, \kappa, \sigma) = \tilde{U}^{\star}_{\text{Td}}(Q, \kappa, \sigma),$$
$$\forall \sigma > 0, \quad (6.57)$$

which leads to the following key insight.

Insight 6.2 *Given the consumption mechanism κ, under the MNO's optimal pricing strategies, both the MNO and the MU can benefit from the supply flexibility. More precisely, it is the buying service that really matters, while the pure selling service adoption cannot change the MNO's revenue or the MUs' payoffs.*

Figure 6.4 shows how the supply flexibility and the consumption flexibility affect MNO revenue and MU payoff under different completion ratios. According to our previous analysis, selling service will not affect the outcome of the equilibrium between MNO and MUs. For example, case (T, Td) and case (T, BY) lead to the same outcome, thus we use a single solid square curve to represent the two cases. The other three curses in each sub-figure follow the similar idea.

- When MUs are not allowed to buy extra data, i.e., $\eta \in \{\text{No}, \text{S1}\}$, the dash curves show that the completion ratio σ in the market does not affect the MNO's revenue and the MU payoff. Comparing the circle curves and the triangle curves shows that rollover mechanism R can always increase the MNO revenue and MU payoff for any completion ratio $\sigma \in [0, 1]$.
- When MUs are allowed to buy extra data, i.e., $\eta \in \{\text{BY}, \text{Td}\}$, the solid curves show that both the MNO revenue and the MU payoff increase in the completion ratio σ. Comparing the diamond curves and the square curves shows that rollover

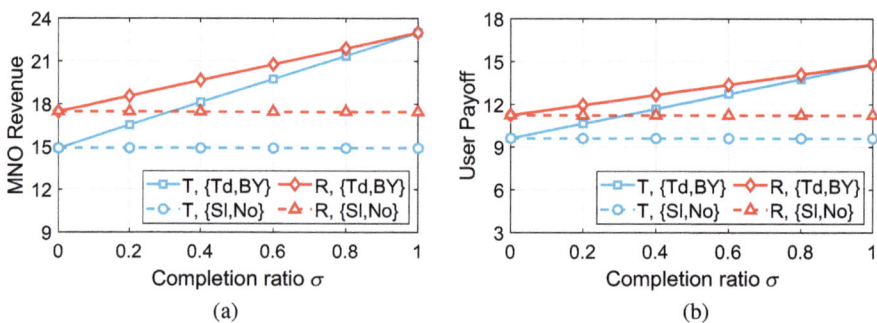

Fig. 6.4 Impact of consumption flexibility and supply flexibility. (**a**) MNO revenue. (**b**) User payoff

mechanism R can increase the MNO revenue and MU payoff if the completion ratio $\sigma < 1$. In other words, the consumption flexibility cannot generate additional benefit when the mobile data market is flexible enough, in the sense that MUs are able to trade and consume data optimally.

6.4 Discussion and Extension

Although this chapter focuses on the consumption flexibility and the supply flexibility primarily in the telecommunication market, our results are also meaningful in *inventory management*. In this chapter, we take the classic newsvendor problem as a representative example to demonstrate it.

As shown in Fig. 6.5a, the classic newsvendor problem considers a supply chain consisting of a manufacturer, a retailer, and the consumers. The retailer purchases the product (e.g., newspaper) from the manufacturer through the regular order, then sells the product to the consumers in the retail market. This operation happens in every *selling season* of the product (e.g., every day for newspaper). The consumers have random demand on the product and the unsold products will be worthless at the end of the selling season. This forces the profit-maximizing retailer to carefully manage its wholesale order. There have been many studies on this classic problems. [63] and [64] provide two comprehensive survey.

The economic interactions in the telecommunication market can be viewed as a special newsvendor problem. Specifically, the MNO is the manufacturer of mobile data with a one-month selling season. The MUs are the retailers who purchase data at the beginning of the selling season, then obtain utility from consuming data. Next we elaborate how the rolling and trading services reflect in the newsvendor problem.

First, the MNO's data *trading* service in the wireless data market corresponds to the supply flexibility. It is related to the buy-back contract and the emergency order in the newsvendor problem.

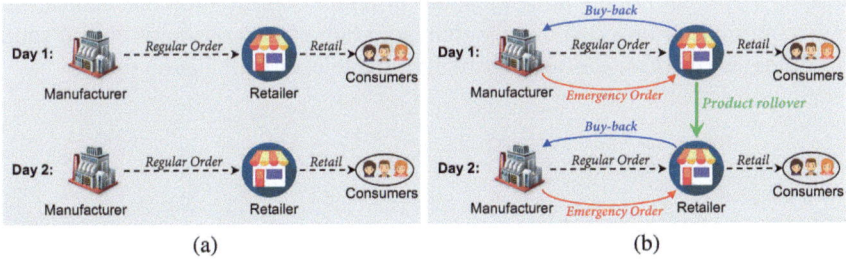

Fig. 6.5 Illustration of different newsvendor problems. (**a**) Classic newsvendor problem. (**b**) Novel newsvendor problem

- *Buy-Back Contract* (Blue arrows in Fig. 6.5b): [65] first studied the buy-back contract, where the manufacturer agrees to buy back the unsold product (from the retailer) at a buy-back price.
- *Emergency Order* (Red arrows in Fig. 6.5b): [66] first examined an inventory system with the emergency order, where the emergency order lead time is assumed negligible. [67] studied the lead time impact of the emergency order.

Second, the MNO's data *rolling* service in the wireless data market increases the mobile users' consumption flexibility. In the newsvendor problem, the manufacturer can also slightly adjust its product and offer the retailer more consumption flexibility (also called *retailing flexibility*). For example, the newspaper manufacturer can add some time-insensitive contents (e.g., serial stories) into the newspapers, besides the time-sensitive news. In that case, the unsold products in the previous selling season still stand in the current selling season, shown by the green arrow in Fig. 6.5b. Hence the retailer enjoys more retailing flexibility comparing with the classic time-sensitive product.

As far as we know, the consumption flexibility (of rolling) has been overlooked in inventory management. We are the first to study the interrelationship between the supply flexibility (of buy-back contract and emergency order) and consumption flexibility (of product rollover).

6.5 Summary

In this chapter, we studied how MNOs can design flexible monthly wireless data contract to better match the MUs' stochastic demand. We proposed to integrate data rolling and trading services, which increase the consumption flexibility and the supply flexibility of the traditional contract, respectively. To analyze the potential trade-off of the rolling and trading services, we studied the strategic interactions between the MNO and MUs in a dynamic game setting. We analyzed the MUs' subscription, data consumption, and data trading behaviors, and derived the MNO's optimal pricing strategy under different consumption mechanisms and supply patterns. We unveiled that the joint adoption of the consumption and supply flexibilities can increase the social benefit of the telecommunication market.

Chapter 7
Conclusion and Outlook

Abstract In recent years, the telecommunications industry has witnessed a paradigm shift toward multi-dimensional flexibility in mobile data service offerings. This trend extends beyond simple pricing adjustments to encompass fundamental restructuring of service architectures across temporal, spatial, and identity-based dimensions. Within this evolving landscape, MNOs face both unprecedented opportunities and complex challenges in redesigning their service portfolios to meet sophisticated consumer demands while maintaining profitable operations.

Keywords General competitive market · Multi-dimensional flexibility · Bounded rationality

7.1 Conclusion

In recent years, the telecommunications industry has witnessed a paradigm shift toward multi-dimensional flexibility in mobile data service offerings. This trend extends beyond simple pricing adjustments to encompass fundamental restructuring of service architectures across temporal, spatial, and identity-based dimensions. Within this evolving landscape, MNOs face both unprecedented opportunities and complex challenges in redesigning their service portfolios to meet sophisticated consumer demands while maintaining profitable operations.

This book provides a comprehensive framework for MNOs to strategically implement three critical flexibility dimensions: (1) temporal flexibility through rollover data mechanisms, (2) location flexibility via innovative day-pass structures, and (3) user-identity flexibility enabled by peer-to-peer data trading platforms. Our systematic examination begins with time-based flexibility, where we establish a dual-market analysis comparing monopoly versus competitive environments. Through rigorous modeling, we demonstrate how rollover mechanisms create distinct value propositions in these differing market structures, particularly in terms of customer retention and revenue optimization.

The investigation then progresses to spatial flexibility solutions, with special emphasis on overseas usage scenarios. We develop a novel decision framework that characterizes MU's optimal strategies when navigating international roaming options, incorporating factors such as travel duration, data consumption patterns, and alternative connectivity solutions. This analysis reveals how day-pass services can simultaneously enhance user satisfaction and operator profitability in global mobility contexts. Our research uncovers the complex interdependencies between temporal and identity flexibility, particularly how rollover mechanisms influence secondary market dynamics. The findings suggest that properly designed trading platforms can create synergistic value when integrated with time-flexible data plans.

Through the above multi-dimensional exploration, our work makes two seminal contributions. First, we quantify the substantial economic impacts of flexible data services across various market configurations, providing empirical evidence of their value creation potential. Second, we establish a foundational taxonomy for understanding flexibility dimensions that is already stimulating new research directions in telecommunications economics. The insights generated have immediate practical implications for service design and long-term theoretical significance for understanding mobile data service evolution.

7.2 Outlook

While preliminary work has established foundational principles for flexible data service provision, the field remains in its nascent stages with numerous compelling research opportunities.

7.2.1 General Competitive Market

In contemporary mobile data markets, oligopolistic competition arises when a limited number of MNOs engage in strategic interactions, each aiming to obtain more market share through differentiated pricing, service quality, and innovative data plans. Unlike perfectly competitive markets, this environment is characterized by interdependent decision-making, where each operator's choices-such as pricing strategies, rollover policies, or quality-of-service (QoS) investments-directly influence competitors' actions and overall market equilibrium. The complexity is further amplified by the need to account for heterogeneous consumer behaviors, regulatory constraints, and rapidly evolving technological capabilities.

To thrive in such a setting, MNOs must design multi-dimensional capability contracts (multi-cap designs) that not only cater to diverse user preferences but also mitigate information asymmetry through incentive-compatible mechanisms. These contracts must incorporate dynamic features such as tiered data allowances, time-flexible rollover options, and peer-to-peer trading capabilities-all while ensuring that users truthfully reveal their private preferences (e.g., usage patterns, willingness to pay). Analytically, this necessitates a game-theoretic framework capable of modeling Bertrand-Nash price competition under differentiated services, while also characterizing equilibrium existence, uniqueness, and stability. Such a framework must integrate Bayesian games to handle incomplete information, dynamic optimization for multi-period interactions, and mechanism design principles to align MNO profitability with consumer welfare. Ultimately, this approach allows us to derive actionable insights into how oligopolistic competition shapes market outcomes, including pricing structures, service innovation, and overall economic efficiency.

Moreover, the equilibrium analysis must consider strategic complementarities (e.g., when one MNO's QoS investment prompts rivals to follow suit) and substitutability (e.g., price undercutting in saturated markets). By incorporating reaction functions and best-response dynamics, we can assess how equilibria evolve under varying market conditions-such as shifts in demand elasticity or regulatory interventions. This not only refines our understanding of real-world competition but also provides MNOs with empirically grounded strategies for optimizing contract designs in an increasingly interconnected and data-driven marketplace.

7.2.2 Multi-Dimensional Flexibility

MNOs are increasingly transforming their three-part tariff structures by incorporating multi-dimensional flexibility across three critical domains: (1) temporal flexibility through dynamic rollover policies that adapt to usage patterns across billing cycles, (2) spatial flexibility enabling seamless cross-border connectivity through innovative roaming packages, and (3) user-identity flexibility facilitated by peer-to-peer data trading platforms. This paradigm shift from rigid, one-size-fits-all data plans to adaptive, multi-attribute service offerings reflects the growing sophistication of consumer demand and the need for more personalized connectivity solutions. The integration of these flexibility dimensions creates a complex optimization challenge, as adjustments in one domain (e.g., extending rollover periods) may significantly impact the effectiveness of others (e.g., roaming package uptake or secondary market liquidity).

Developing an effective implementation framework requires addressing several interconnected challenges. For spatial flexibility in particular, MNOs must establish novel inter-operator settlement mechanisms that fairly distribute revenues from roaming usage while maintaining incentives for infrastructure investment. This necessitates both technical innovations in real-time usage tracking across partner networks and economic innovations in dynamic pricing models that account for fluctuating demand patterns in different geographical markets. The framework must also resolve the inherent tension between competitive differentiation and inter-operability standards, ensuring that flexibility enhancements do not inadvertently create new market barriers. From an analytical perspective, this calls for advanced optimization techniques that can simultaneously model stochastic demand across multiple dimensions, contractual constraints, and strategic interactions between market participants, while quantifying the trade-offs between short-term revenue maximization and long-term customer retention. Ultimately, the success of these multi-dimensional tariff structures hinges on finding equilibrium solutions that balance technical feasibility, economic viability, and regulatory compliance across diverse operating environments.

7.2.3 Bounded Rationality

A growing body of empirical research in behavioral telecommunications economics reveals that MUs systematically deviate from neoclassical rational choice models when making data consumption decisions. These deviations stem from fundamental cognitive limitations—including finite attention spans, computational constraints, and heuristic-driven decision making—which lead to bounded rationality in practice. Usage pattern analyses consistently demonstrate that when faced with complex tariff structures involving multiple dimensions (data allowances, rollover policies, speed tiers), users overwhelmingly resort to satisficing behaviors rather than utility-maximizing strategies. This manifests in several observable ways: heavy reliance on default options regardless of optimality, disproportionate sensitivity to salient but potentially irrelevant plan features, and systematic underestimation of future data needs due to present bias.

These robust behavioral findings necessitate a paradigm shift in mobile service design, moving beyond traditional models that presume perfect information processing and unlimited cognitive capacity. Effective redesigns must incorporate three evidence-based principles: First, simplified choice architectures should strategically limit options to 3–5 meaningful alternatives while eliminating redundant dimensions that create analysis paralysis. Second, behaviorally-informed default options must be calibrated to satisficing thresholds identified through experimental testing. Third, active decision supports like real-time usage predictors and personalized plan comparators can help bridge users' cognitive gaps without removing agency. Implementing these principles requires fundamentally rethinking optimization objectives from maximizing theoretical ideal usage to minimizing behavioral frictions while preserving sufficient flexibility. The resulting frameworks must account for the homogeneity of user comprehension, treating decision-making capability as a design variable rather than a fixed constraint. This represents a significant departure from conventional network economics models, with important implications for pricing strategy, service differentiation, and regulatory policy in increasingly complex digital service markets.

References

1. Ericsson Mobility Report, *Mobile network data traffic* (2024) [Online]. Available: https://www.ericsson.com/49ed78/assets/local/reports-papers/mobility-report/documents/2024/ericsson-mobility-report-june-2024.pdf
2. uSwitch, *UK Wastes Almost 150 Million Gigabytes Of Mobile Data Every Month* (2018)
3. AT&T, *AT&T Rollover data FAQs* (2023) [Online]. Available: https://www.att.com/support/smallbusiness/article/smb-wireless/KM1205715
4. China Mobile, *China Mobile Rollover Data* (2024) [Online]. Available: https://www.10086.cn/4G/zixuantc/hb/
5. Sky Mobile, *Sky Mobile Rollover data* (2024) [Online]. Available: https://www.att.com/support/smallbusiness/article/smb-wireless/KM1205715
6. AT&T, *AT & T International Day Pass* (2024) https://www.att.com/offers/international-plans/day-pass.html
7. SMARTY, *SMARTY's Data Discount Explained: Money Back for Unused Data* (2024) [Online]. Available: https://www.simsherpa.com/networks/smarty-mobile/how-data-discount-works
8. China Mobile, CMHK *2nd Exchange Market* (2024) [Online]. Available: https://www.hk.chinamobile.com/tc/2cm_intro.html
9. S. Sen, C. Joe-Wong, S. Ha, M. Chiang, Incentivizing time-shifting of data: a survey of time-dependent pricing for internet access. IEEE Commun. Mag. **50**(11) (2012)
10. S. Sen, C. Joe-Wong, S. Ha, M. Chiang, A survey of smart data pricing: Past proposals, current plans, and future trends. ACM Comput. Surv. (CSUR) **46**(2), 15 (2013)
11. R.B. Wilson, *Nonlinear Pricing* (Oxford University Press on Demand, Oxford, 1993)
12. A. Lambrecht, K. Seim, B. Skiera, Does uncertainty matter? consumer behavior under three-part tariffs. Marketing Sci. **26**(5), 698–710 (2007)
13. L. Xu, J. Duan, Y.J. Hu, Y. Cheng, Y. Zhu, Forward-looking behavior in mobile data consumption and targeted promotion design: a dynamic structural model. SSRN: https://ssrn.com/abstract=2540533
14. A. Nevo, J.L. Turner, J.W. Williams, Usage-based pricing and demand for residential broadband. Econometrica **84**(2), 411–443 (2016)
15. G. Fibich, R. Klein, O. Koenigsberg, E. Muller, Optimal three-part tariff plans. Oper. Res. **65**(5), 1115–1428 (2017)
16. H.K. Bhargava, M. Gangwar, On the optimality of three-part tariff plans: when does free allowance matter? Oper. Res. **66**(6), 1517–1532 (2018)
17. J.F. Wu, B. Behzad, Optimal three-part tariff pricing with spence-mirrlees reservation prices. Math. Methods Oper. Res. **97**(2), 233–258 (2023)

18. L. Zheng, C. Joe-Wong, Understanding rollover data, in *IEEE Conference on Computer Communications Workshops (INFOCOM WKSHPS)* (2016)
19. Z. Wang, L. Gao, J. Huang, Pricing optimization of rollover data plan, in *15th International Symposium on Modeling and Optimization in Mobile, Ad Hoc, and Wireless Networks (WiOpt)* (2017)
20. Y. Wei, J. Yu, T.M. Lok, L. Gao, A novel mobile data contract design with time flexibility. Preprint (2018). arXiv:1806.07308
21. L. Zheng, C. Joe-Wong, C.W. Tan, S. Ha, M. Chiang, Customized data plans for mobile users: feasibility and benefits of data trading. IEEE J. Selected Areas Commun. **35**(4), 949–963 (2017)
22. J. Yu, M.H. Cheung, J. Huang, H.V. Poor, Mobile data trading: behavioral economics analysis and algorithm design. IEEE J. Selected Areas Commun. **35**(4), 994–1005 (2017)
23. W. Huang, Y. Li, Y. Chen, Overage disutility, user trading, and tariff design. Production Oper. Manag. **30**(10), 3758–3783 (2021)
24. Q. Ma, Y.-F. Liu, J. Huang, Time and location aware mobile data pricing. IEEE Trans. Mobile Comput. **15**(10), 2599–2613 (2015)
25. P. Maille, B. Tuffin, Enforcing free roaming among eu countries: an economic analysis, in *2017 13th International Conference on Network and Service Management (CNSM)* (IEEE, 2017), pp. 1–4
26. L. Duan, J. Huang, B. Shou, Pricing for local and global wi-fi markets IEEE Trans. Mobile Comput. **14**(5), 1056–1070 (2014)
27. F. Wang, L. Duan, J. Niu, Optimal pricing of user-initiated data-plan sharing in a roaming market. IEEE Trans. Wirel. Commun. **17**(9), 5929–5944 (2018)
28. M. Zhang, L. Gao, J. Huang, M.L. Honig, Hybrid pricing for mobile collaborative internet access. IEEE/ACM Trans. Netw. **27**(3), 986–999 (2019)
29. Z. Wang, L. Gao, J. Huang, Exploring time flexibility in wireless data plans. IEEE Trans. Mobile Comput. **18**(9), 2048–2061 (2018)
30. Z. Wang, L. Gao, J. Huang, Duopoly competition for mobile data plans with time flexibility. IEEE Trans. Mobile Comput. **19**(6), 1286–1298 (2019)
31. Z. Wang, L. Gao, J. Huang, B. Shou, Economic viability of data trading with rollover, in *IEEE International Conference on Computer Communications (INFOCOM)* (2019)
32. Z. Wang, L. Gao, J. Huang, Multi-dimensional contract design for mobile data plan with time flexibility, in *ACM International Symposium on Mobile Ad Hoc Networking and Computing (MobiHoc)* (2018), pp. 51–60
33. Z. Wang, L. Gao, J. Huang, Multi-cap optimization for wireless data plans with time flexibility. IEEE Trans. Mobile Comput. **19**(9), 2145–2159 (2019)
34. Z. Wang, L. Gao, J. Huang, Travel with your mobile data plan: a location-flexible data service, in *IEEE Conference on Computer Communications (INFOCOM)* (2020)
35. Z. Wang, L. Gao, J. Huang, Location-flexible mobile data service in overseas market. IEEE Trans. Mobile Comput. **21**(2), 629–643 (2020)
36. Z. Wang, L. Gao, J. Huang, B. Shou, Toward flexible wireless data services. IEEE Commun. Mag. **57**(12), 25–30 (2019)
37. Z. Wang, L. Gao, T. Wang, J. Luo, Monetizing edge service in mobile internet ecosystem. IEEE Trans. Mobile Comput. **21**(5), 1751–1765 (2020)
38. R.T. Ma, Usage-based pricing and competition in congestible network service markets. IEEE/ACM Trans. Netw. **24**(5), 3084–3097 (2016)
39. S. Sen, C. Joe-Wong, S. Ha, The economics of shared data plans, in *Annual Workshop on Information Technologies and Systems* (2012)
40. R.H. Frank, A.J. Glass, *Microeconomics and Behavior* (McGraw-Hill, New York, 1991)
41. CapEx, *Capital Expenditure* (2024) [Online]. Available: https://en.wikipedia.org/wiki/Capital_expenditure
42. W.P. Rogerson, E. Charles, The economics of data caps and free data services in mobile broadband. CTIA Report, 2016

References

43. D.A. Lyons, The impact of data caps and other forms of usage-based pricing for broadband access. *Mercatus Center Working Papers* (2012)
44. OpEx, *Operating Expense*. [Online]. Available: https://en.wikipedia.org/wiki/Operating_expense
45. P. Nabipay, A. Odlyzko, Z.-L. Zhang, Flat versus metered rates, bundling, and bandwidth hogs, in *6th Workshop on the Economics of Networks, Systems, and Computation* (2011)
46. R.F. Bass, *Stochastic Processes* (Cambridge University Press, Cambridge, 2011)
47. Failure Rate, (2024) [Online]. Available: https://en.wikipedia.org/wiki/Failure_rate
48. X. Brusset, Properties of distributions with increasing failure rate. MPRA Paper 18299, University Library of Munich (2009)
49. D. Fudenberg, J. Tirole, *Game Theory* (The MIT Press, Cambridge, 1991)
50. W. Dai, S. Jordan, The effect of data caps upon isp service tier design and users. ACM Trans. Internet Technol. **15**(2), 8 (2015)
51. X. Wang, R.T. Ma, Y. Xu, On optimal two-sided pricing of congested networks. Proc. ACM Meas. Anal. Comput. Syst. **1**(1), 7 (2017)
52. L. Gao, X. Wang, Y. Xu, Q. Zhang, Spectrum trading in cognitive radio networks: A contract-theoretic modeling approach. IEEE J. Selected Areas Commun. **29**(4), 843–855 (2011)
53. L. Duan, L. Gao, J. Huang, Cooperative spectrum sharing: a contract-based approach. IEEE Trans. Mobile Comput. **13**(1) (2014)
54. Y. Zhang, L. Song, W. Saad, Z. Dawy, Z. Han, Contract-based incentive mechanisms for device-to-device communications in cellular networks. IEEE J. Selected Areas Commun. **33**(10), 2144–2155 (2015)
55. Z. Wang, L. Gao, J. Huang, Multi-dimensional contract design for mobile data plan with time flexibility, in *International Symposium on Mobile Ad Hoc Networking and Computing (MobiHoc)* (2018)
56. R. Bellman, *Dynamic Programming* (Courier Corporation, Chelmsford, 2013)
57. X. Wang, R.T. Ma, Y. Xu, The role of data cap in optimal two-part network pricing. IEEE/ACM Trans. Netw. **25**(6), 3602–3615 (2017)
58. M.J. Neely, Stochastic network optimization with application to communication and queueing systems. Synth. Lect. Commun. Networks. **3**(1), 1–211 (2010)
59. E. Hazan, et al., Introduction to online convex optimization. Found. Trends Optim. **2**(3–4), 157–325 (2016)
60. Y. Jin, Z. Pang, Smart data pricing: the value of shared data plans. Service Sci. **8**(4), 386–405 (2016)
61. S. Shalev-Shwartz, Online learning and online convex optimization. Found. Trends Mach. Learn. **4**(2), 107–194 (2012)
62. E. Maskin, J. Riley, Monopoly with incomplete information. The RAND J. Econ. **15**(2), 171–196 (1984)
63. M. Khouja, The single-period (news-vendor) problem: literature review and suggestions for future research. Omega **27**(5), 537–553 (1999)
64. Y. Qin, R. Wang, A.J. Vakharia, Y. Chen, M.M. Seref, The newsvendor problem: review and directions for future research. Eur. J. Oper. Res. **213**(2), 361–374 (2011)
65. B.A. Pasternack, Optimal pricing and return policies for perishable commodities. Marketing Sci. **4**(2), 166–176 (1985)
66. M. Rosenshine, D. Obee, Analysis of a standing order inventory system with emergency orders. Oper. Res. **24**(6), 1143–1155 (1976)
67. C. Chiang, G.J. Gutierrez, A periodic review inventory system with two supply modes. Eur. J. Oper. Res. **94**(3), 527–547 (1996)

MIX
Papier aus verantwortungsvollen Quellen
Paper from responsible sources
FSC® C105338

If you have any concerns about our products,
you can contact us on
ProductSafety@springernature.com

In case Publisher is established outside the EU,
the EU authorized representative is:
**Springer Nature Customer Service Center GmbH
Europaplatz 3, 69115 Heidelberg, Germany**

Printed by Libri Plureos GmbH
in Hamburg, Germany